Learn Arduino Prototyping in 10 days

Your crash course to build innovative devices

Kallol Bosu Roy Choudhuri

BIRMINGHAM - MUMBAI

Learn Arduino Prototyping in 10 days

First published: June 2017

Production reference: 1270617

Published by Packt Publishing Ltd.
Livery Place
35 Livery Street
Birmingham
B3 2PB, UK.

ISBN 978-1-78829-068-5

www.packtpub.com

Credits

Author
Kallol Bosu Roy Choudhuri

Reviewers
Aaron Srivastava
Fangzhou Xia

Commissioning Editor
Vijin Boricha

Acquisition Editor
Divya Poojari

Content Development Editor
Deepti Thore

Technical Editor
Deepti Tuscano

Copy Editor
Safis Editing

Project Coordinator
Shweta H Birwatkar

Proofreader
Safis Editing

Indexer
Mariammal Chettiyar

Graphics
Tania Dutta

Production Coordinator
Nilesh Mohite

About the Author

Kallol Bosu Roy Choudhuri is a passionate Arduino prototype creator. He is a professional computer science engineer, an IT enterprise architect, and thought leader with several years of industry experience. He provides expert consultancy on matters of software architecture and works on many exciting and innovative IoT devices and solutions.

The author comes with more than 14 years of industry experience and works passionately with an entrepreneurial spirit. He practices as a senior architect in the Enterprise Architecture Group at Atos India, and works on many exciting and innovative engagements.

He specializes in architecture and design, performance engineering, legacy modernization, IT transformations, assessments, strategy, presales, and consulting. Prior to Atos, he has worked for other multinational software companies such as Tata Consultancy Services and Cognizant Technology Solutions and has served many Fortune 500 customers at their site offices in Europe and North America.

I would like to thank my wife, daughter, family, and close friends who have helped me in completing this book.

About the Reviewers

Aaron Srivastava is a software engineer for Fujifilm Medical Systems. He has years of experience building Arduino projects and actively looks for new projects to implement.

Fangzhou Xia is currently a PhD student in the **Mechanical Engineering** (**MEng**) department at **Massachusetts Institute of Technology** (**MIT**). He received his ME bachelor's degree at **University of Michigan** (**UM**) and **Electrical and Computer Engineering** (**ECE**) bachelor's degree at **Shanghai Jiao Tong University** (**SJTU**). His areas of interest in mechanical engineering include system control, robotics, product design, and manufacturing automation. His areas of interest in electrical engineering and computer science include web application development, embedded system implementation, data acquisition system setup, and machine learning applications.

www.PacktPub.com

For support files and downloads related to your book, please visit www.PacktPub.com.

Did you know that Packt offers eBook versions of every book published, with PDF and ePub files available? You can upgrade to the eBook version at www.PacktPub.com and as a print book customer, you are entitled to a discount on the eBook copy. Get in touch with us at service@packtpub.com for more details.

At www.PacktPub.com, you can also read a collection of free technical articles, sign up for a range of free newsletters and receive exclusive discounts and offers on Packt books and eBooks.

https://www.packtpub.com/mapt

Get the most in-demand software skills with Mapt. Mapt gives you full access to all Packt books and video courses, as well as industry-leading tools to help you plan your personal development and advance your career.

Why subscribe?

- Fully searchable across every book published by Packt
- Copy and paste, print, and bookmark content
- On demand and accessible via a web browser

Customer Feedback

Thanks for purchasing this Packt book. At Packt, quality is at the heart of our editorial process. To help us improve, please leave us an honest review on this book's Amazon page at `https://www.amazon.com/dp/1788290682` .

If you'd like to join our team of regular reviewers, you can e-mail us at `customerreviews@packtpub.com`. We award our regular reviewers with free eBooks and videos in exchange for their valuable feedback. Help us be relentless in improving our products!

Table of Contents

Preface   1

Chapter 1: Boot Camp   9
   Organization of the chapters   9
   How to use the book   10
   Things you will need to get started   11
   Things you will learn in this book   12
   Summary   15

Chapter 2: The Arduino Platform   17
   Introduction to the Arduino platform   18
   Overview of Arduino prototyping   21
   Setting up the Arduino board   23
   Arduino program structure and execution   24
   Understanding the first Arduino sketch   26
   Compiling, loading and running a sketch   28
   Commonly used in-built C sketch functions   31
   Digital input and output   33
   Analog input and output   35
   Try the following   37
   Things to remember   37
   Summary   38

Chapter 3: Day 1 - Building a Simple Prototype   39
   The three LED project   40
   Rationale for using a resistor   40
   The Piezo Buzzer project   47
   Using transistors   51
   Using diodes   55
   LED with a push button   59
   Try the following   63
   Things to remember   63
   Summary   64

Chapter 4: Day 2 - Interfacing with Sensors   65
   Types of sensor components   65
     Basic sensor components   66

Using a basic sensor - photodiode   66
Using a basic sensor - photo resistor (LDR)   71
Using integrated sensor modules   76
Using a temperature sensor module (with an Arduino library)   76
Understanding sensor module datasheets   77
Installing the sensor-specific Arduino library   81
Sensor interfacing sketch   82
Viewing the program output   84
Using a soil moisture sensor module (without an Arduino library)   86
Soil moisture sensor circuit   88
Soil moisture sensor sketch   89
Future inspiration   92
Try the following   93
Things to remember   93
Summary   94

Chapter 5: Day 3 - Building a Compound Device   95
Compound devices   96
Building a smoke detector   96
Smoke detector - Digital I/O method   98
Smoke detector sketch - Digital I/O method   100
Smoke detector (analog I/O method)   104
Smoke detector sketch (analog I/O method)   105
Local storage with SD card modules   109
Try the following   119
Things to remember   121
Summary   122

Chapter 6: Day 4 - Building a Standalone Device   123
Standalone devices   124
External power supply options   125
Determining power source capacity   126
Building a distance measurement device   129
Distance measurement device circuit   132
Distance measurement device sketch   135
Operating the distance measurement device   138
Finishing touches   139
Try the following   142
Things to remember   142
Summary   143

Chapter 7: Day 5 - Using Actuators   145

About actuators 146
Special considerations while using DC motors 146
A basic DC motor prototype 147
Basic DC motor sketch 150
DC motor speed control - PWM method 151
DC motor speed control sketch 154
Using Arduino interrupts 156
Interfacing with a servo motor 158
Servo motor control circuit 159
Servo motor control sketch 161
Future inspiration 164
Try the following 165
Things to remember 165
Summary 166

Chapter 8: Day 6 - Using AC Powered Components 167
Using relays with AC powered devices 168
Part 1 - Simulation of sound activated light bulb controller 171
The sound-activated device sketch 174
Part 2 - Actual prototype for sound activated light bulb controller 177
Future inspiration - Automatic room lights 180
Try the following 181
Things to remember 182
Summary 182

Chapter 9: Day 7 - The World of Transmitters, Receivers, and Transceivers 183
Understanding Infrared communications 183
Infrared communication frequency 184
Infrared communication protocol 185
Hacking into an existing remote control 186
Building an Infrared receiver device 187
The Arduino Infrared library 188
Using IR receiver TSOP series IR receivers 194
Using IR receiver SM0038 200
Building an Infrared transmitter device 201
Using an IR transmitter LED 201
Controlling Arduino projects 205
Transceivers 206
Try the following 207

Things to remember   207
Summary   208

Chapter 10: Day 8 - Short Range Wireless Communications   209
Building a radio frequency device   210
Using the nRF24L01 transceiver module   211
Wiring nRF24L01 with Arduino   212
Downloading the open source RF library for Arduino   214
Transmitting radio frequency waves   215
Receiving radio frequency signals   217
Testing the RF transmitter-receiver pair   219
Bluetooth communications   219
Using the HC-05 Bluetooth module   220
Connecting HC-05 to Arduino Uno   221
HC-05 sketch   222
Communicating with the HC-05 prototype   224
Try the following   225
Things to remember   225
Summary   226

Chapter 11: Day 9 - Long-Range Wireless Communications   227
The GSM module   228
AT commands   230
GSM module interfacing with Arduino Uno   231
GSM module sketch   232
Forest fire early warning system - Inspiration   237
Try the following   240
Things to remember   240
Summary   241

Chapter 12: Day 10 - The Internet of Things   243
Introduction to IOT   244
IoT edge devices   245
IoT Cloud platforms   247
IoT cloud configuration   249
Step 1 - IoT cloud registration   249
Step 2 - Configuring an edge device channel   250
Edge device setup   257
Building the edge device   257
Edge device sketch   259
Smart retail project inspiration   263
IOT project considerations   265

Try the following   266

Things to remember   266

Summary   267

Index   269

Preface

This book has been crafted to serve as a quick crash course for becoming well acquainted with the Arduino platform in just 10 days. The primary focus of the book is to empower the reader to use the Arduino platform by applying basic fundamental principles and be able to apply those fundamental principles for building almost any type of physical device.

The uniqueness of the book lies in its practical approach and pedagogy. This book does not try to explain all the possible projects that can be achieved with the Arduino platform, but instead establishes the fundamental types of projects and techniques using which readers will be able to build any device prototype on their own.

The book is intended to serve as a beginner's crash course for professionals, hobbyists, and students who are tech savvy, have a basic level of C programming knowledge and basic familiarity with electronics, be it for embedded systems or for the Internet of Things. The book introduces some basic electronics concepts and useful programming functions that are essential for use with the Arduino platform. It will save the reader hours of research work, by presenting all the required knowledge in a crisp and concise package--almost everything to get started with in one single place!

While writing this book, great care has been taken to present the fundamental principles in a pragmatic and scientific manner and guide the audience through a graded series of chapters, based on the application of fundamental principles and increasing level of complexity. So by the end of the book, the readers will feel confident about taking on new device prototyping challenges, completely on their own.

What this book covers

The 10-day journey includes various practical aspects of the Arduino platform, presented in a clear and concise manner. Each chapter in this book is intended to correspond to a day's worth of study. Each chapter introduces and demonstrates unique practical fundamentals through hands-on examples; that must be assimilated to become self-reliant on the topic of Arduino prototyping.

Chapter 1, *Boot Camp*, welcomes you with this Boot Camp chapter and proceeds to guide you on how to use the book. It lists the hardware components and devices that must be procured in order to follow through the learning path outlined in the chapters.

Chapter 2, *The Arduino Platform*, introduces you to the Arduino platform. First, we will see what the Arduino platform is all about. Then, there will be quick introductory topics regarding fundamentals for getting started. This will be followed by a quick but in-depth look at the first Arduino code.

Chapter 3, *Day 1 - Building a Simple Prototype*, starts your journey by learning to build a simple device prototype. This will be your first hardware-software integrated prototype. Two easy-to-work prototypes with examples have been chosen for this chapter. The first example will be to emit light patterns. While the second example will be to emit basic sounds and play a musical tone.

Chapter 4, *Day 2 - Interfacing with Sensors*, describes how to work with sensors in general. You might have seen automatic doors that slide open once somebody goes near the door. These automatic systems are usually based on sensors, microcontrollers and embedded software. In this chapter, we will learn the fundamental technique of interfacing with sensors in general.

Chapter 5, *Day 3 - Building a Compound Device*, takes a step ahead by explaining how to build compound device prototypes using the Arduino platform. Compound devices are a very important topic as real-world devices are usually composed of multiple devices integrated with a central microcontroller. This chapter will provide a hands-on example to building a compound device.

Chapter 6, *Day 4 - Building a Standalone Device*, facilitates building real-world device prototypes. Independent power sources, not from a computer's USB port, will be used so that the device prototype can work without being connected to a computer. In this chapter, we will learn how to make standalone devices that have their independent power sources, a power switch, and a container.

Chapter 7, *Day 5 - Using Actuators*, proceeds to work on our first project that uses diodes and transistors with a DC motor (an example of an actuator) powered from an independent battery-based power source. This is an advanced level chapter and is designed with a lot of concepts and components, which builds upon the knowledge gathered so far from the previous chapters.

Chapter 8, *Day 6 - Using AC Powered Components*, presents a unique presentation for introduction to the fundamentals of interfacing and controlling AC-powered electrical devices with the Arduino platform. This topic was specifically chosen for a wholesome completeness of the 10-day crash course.

Chapter 9, *Day 7 - The World of Transmitters, Receivers, and Transceivers*, reveals an exciting chapter on infrared transmitters and receivers. As you read through, this chapter will unravel and demystify some embedded world techniques used for transmitting and receiving data from one device to another using wireless signals.

Chapter 10, *Day 8 - Short Range Wireless Communications*, introduces us to hands-on techniques used for transmitting and receiving data from one device to another using wireless radio signals (RF and Bluetooth). We will start learning wireless communications using Radio Frequency (RF).

Chapter 11, *Day 9 - Long Range Wireless Communications*, provides a fascinating introduction to the exciting world of telephony. In this chapter, we will learn how to use a GSM module with the Arduino platform.

Chapter 12, *Day 10 - The Internet of Things*, explains how to use the Arduino platform in the fast emerging Internet of Things world. All of us have heard about the buzzword IoT (Internet of Things). The Internet of Things is a growing network of physical devices that can connect to the existing Internet and exchange data with other devices.

All the chapters have working Arduino code and circuit building specifications and instructions.

What you need for this book

All the examples in this book use the Arduino Uno R3 platform. This version of the Arduino board was chosen because it is the most often recommended microcontroller board for learning hardware/software prototyping. Once the techniques have been mastered, the reader will be able to adapt the examples to other development boards and devices as well.

It is expected that the reader possesses the following basic skills that will be required for engaging in this 10-day Arduino prototyping crash course:

- Basic knowledge of C programming (simple variables, functions, if statements, for loops and functions will suffice, nothing fancy is needed) is required to follow through the chapters in this book.
- Familiarity with basic electronic components (resistors, diodes, transistors, breadboards, circuits, and so on). You do not need to know the fundamentals of how these components work; the book will explain everything ground up starting from fundamental concepts.

Each chapter in this book uses many hardware components and has a list of hardware parts required to build the example prototypes in the chapter. In order to provide you with a comprehensive experience; a concise list of the hardware components that you will need, has been provided in `Chapter 1`, *Boot Camp*.

Who this book is for

Arduino prototyping typically demands two general skills. The first skill is familiarity with basic C programming, while the second aspect is familiarity with electronic components. Familiarity with C language is required because the Arduino programs (known as Sketches) are written using C. While building the Arduino device prototypes will require some familiarity with basic electronics components.

This book has been written in a balanced manner. The book is for readers who know basic C programming (variables, functions, if statements, and for loops will suffice), and want to quickly jump start building device prototypes on the Arduino platform, without having to read through time-consuming documentation, tutorials, and tireless research. Additionally, the book assumes that the reader has very little familiarity with electronics and explains everything from scratch. All you need is your "will" to learn and the rest will be taken care by this book!

This book has been designed for the following audience:

- For an enthusiastic DIY (Do-It-Yourself) hobbyist (from schools to colleges to professionals from any walk of life)--basically anyone who wants to learn how to make microcontroller-based electronics devices.
- It can be used by technologists and engineers who want to upskill themselves in a very rapid manner and start working in the field of IoT (Internet of Things) or get introduced to Embedded Systems device prototyping.
- It can be used by university and college students as well as teachers and lecturers as part of their practical lab courseware for microcontroller-based subjects.
- It is also suitable for higher education/library material for embedded systems and/or practical engineering (computer science, electronics, electrical, instrumentation, telecommunication, and allied disciplines) lab courses.

Conventions

In this book, you will find a number of text styles that distinguish between different kinds of information. Here are some examples of these styles and an explanation of their meaning.

Code words in text, database table names, folder names, filenames, file extensions, pathnames, dummy URLs, user input, and Twitter handles are shown as follows: "The `setup()` function runs only once every time the board is either reset or powered up."

A block of code is set as follows:

```
// Signal a quick LOW just before giving a HIGH signal
digitalWrite(3, LOW);
delayMicroseconds(2);

// After 2 micro-seconds of LOW signal, give a HIGH signal
// to trigger the sensor
digitalWrite(3, HIGH);
// Keep the digital signal HIGH forat least 10 micro-seconds
// (required by HC-SR04 to activate emission of ultra-sonic
// waves)
delayMicroseconds(10);

// After 10 micro-seconds, send a LOW signal
digitalWrite(3, LOW);
```

Warnings or important notes appear in a box like this.

Tips and tricks appear like this.

Reader feedback

Feedback from our readers is always welcome. Let us know what you think about this book-what you liked or disliked. Reader feedback is important for us as it helps us develop titles that you will really get the most out of.

To send us general feedback, simply e-mail feedback@packtpub.com, and mention the book's title in the subject of your message.

If there is a topic that you have expertise in and you are interested in either writing or contributing to a book, see our author guide at www.packtpub.com/authors.

Customer support

Now that you are the proud owner of a Packt book, we have a number of things to help you to get the most from your purchase.

Downloading the example code

You can download the example code files for this book from your account at http://www.p acktpub.com. If you purchased this book elsewhere, you can visit http://www.packtpub.c om/support and register to have the files e-mailed directly to you.

You can download the code files by following these steps:

1. Log in or register to our website using your e-mail address and password.
2. Hover the mouse pointer on the **SUPPORT** tab at the top.
3. Click on **Code Downloads & Errata**.
4. Enter the name of the book in the **Search** box.
5. Select the book for which you're looking to download the code files.
6. Choose from the drop-down menu where you purchased this book from.
7. Click on **Code Download**.

Once the file is downloaded, please make sure that you unzip or extract the folder using the latest version of:

- WinRAR / 7-Zip for Windows
- Zipeg / iZip / UnRarX for Mac
- 7-Zip / PeaZip for Linux

The code bundle for the book is also hosted on GitHub at https://github.com/PacktPubl ishing/Learn-Arduino-Prototyping-in-10-days. We also have other code bundles from our rich catalog of books and videos available at https://github.com/PacktPublishing/. Check them out!

Downloading the color images of this book

We also provide you with a PDF file that has color images of the screenshots/diagrams used in this book. The color images will help you better understand the changes in the output. You can download this file from https://www.packtpub.com/sites/default/files/down loads/LearnArduinoPrototypingin10days_ColorImages.pdf.

Errata

Although we have taken every care to ensure the accuracy of our content, mistakes do happen. If you find a mistake in one of our books-maybe a mistake in the text or the code-we would be grateful if you could report this to us. By doing so, you can save other readers from frustration and help us improve subsequent versions of this book. If you find any errata, please report them by visiting http://www.packtpub.com/submit-errata, selecting your book, clicking on the **Errata Submission Form** link, and entering the details of your errata. Once your errata are verified, your submission will be accepted and the errata will be uploaded to our website or added to any list of existing errata under the Errata section of that title.

To view the previously submitted errata, go to https://www.packtpub.com/books/content/support and enter the name of the book in the search field. The required information will appear under the **Errata** section.

Piracy

Piracy of copyrighted material on the Internet is an ongoing problem across all media. At Packt, we take the protection of our copyright and licenses very seriously. If you come across any illegal copies of our works in any form on the Internet, please provide us with the location address or website name immediately so that we can pursue a remedy.

Please contact us at copyright@packtpub.com with a link to the suspected pirated material.

We appreciate your help in protecting our authors and our ability to bring you valuable content.

Questions

If you have a problem with any aspect of this book, you can contact us at questions@packtpub.com, and we will do our best to address the problem.

1
Boot Camp

"For all things magic and beyond, there is the magic wand. For us lesser mortals, we use the Arduino!"
- Kallol Bosu Roy Choudhuri

The book welcomes you with this *Boot Camp* chapter that explains the organization of the chapters in this book, followed by guidance on how to use the book for maximum benefit. It lists the hardware components and devices that must be procured in order to follow the learning path outlined in the chapters. Most importantly, you will find a comprehensive list of the fundamentals, concepts, and projects that you will learn from this book.

Organization of the chapters

The chapters in this book are dedicated to laying down a strong technical foundation for the years to come in your Arduino-based device prototyping journey. To start off, we will see what the Arduino platform is and then through a series of ten, carefully graded chapters, in order of increasing level of complexity, we will become familiar with building device prototypes based on the Arduino platform.

The following diagram depicts the broad areas that the reader will learn in this book:

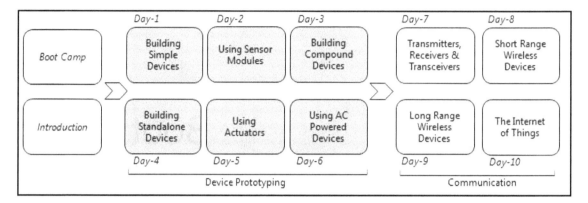

Figure 1: Organization of chapters in the book

At the end of each chapter, there is a *Try the following* section, meant to inspire the reader's imagination and take them a step nearer to becoming more self-reliant. The reader is urged to try out the suggestions in this section to get further insights into the various aspects covered in the chapters.

 All the code files used in the chapters are available at the following online location:
https://github.com/PacktPublishing/Learn-Arduino-Prototyping-in-10-Days.

At the very end of each chapter there is a *Things to remember* section. This section lists all the important points that were covered and is meant to serve as a quick recollection of the concepts learnt in a chapter.

How to use the book

The initial chapters will provide detailed steps for building the example prototypes. However, sometimes during the course of the chapters, the reader will be subtly encouraged to become more independent. As you proceed through the chapters you will notice that the book starts covering only the bare minimum details, while you start understanding and doing more on your own.

- If you are a beginner, go through the chapters one by one, in the order laid out in the book.

- If you have a basic knowledge of Arduino prototyping then you may start from Chapter 4, *Day 2 - Interfacing with Sensors*.
- If you are already aware of prototyping with Arduino, then you may start from Chapter 6, *Day 4 - Building a Standalone Device,* from Day 5 onwards.
- Chapters up to Day 6 are for general device prototyping, whereas starting from Day 7 onwards the chapters deal with wireless communications. You may choose as per your requirements.

If the learning curve is perceived to be steep, then you are advised to pause, relax, and revisit the chapters as many times as needed. However, it is recommended that at no stage of the book should you rush; imbibe steadily and let the concepts sink into you so that you can understand and feel how Arduino-based prototyping is done.

Things you will need to get started

It is advisable to purchase an Arduino Uno starter kit to begin with. Usually a starter kit should cover most of the basic hardware components. However, there may be some advanced components (such as the sensor modules) that will have to be purchased separately. It is recommended you perform your own research for procuring the components at your convenience.

The components in the following table have been listed uniquely and incrementally. For example, a Piezo Buzzer has been used on Day 1 as well as Day 3, however, it has been listed only once for Day 1. This has been done intentionally to assist you during your procurement phase. However, if you want to focus on a particular chapter then you are advised to visit that chapter for a complete list of hardware parts required for the prototypes in that chapter:

Chapter	Hardware requirements (mentioned incrementally)
The Arduino platform	Arduino Uno R3 board, USB A to USB B cable
Day 1	One half-sized breadboard, three red LEDs, three pieces 220 Ohms resistors, a jumper wire set (male-to-male, male-to-female, female-to-female), one Piezo Buzzer, one piece 100 Ohms resistor, one N2222 transistor, two pieces 150 Ohms resistors, one piece IN4007 diode, one push button, one piece 10K Ohms resistor

Chapter	Hardware requirements (mentioned incrementally)
Day 2	One photodiode, one photo resistor (LDR), one DHT11 temperature sensor, one 5K Ohms resistor, one soil moisture sensor
Day 3	One MQ2 gas sensor, one SD card module (micro SD), one micro SD card
Day 4	One full-sized breadboard, one piece 9V battery, one HC-SR04 ultrasonic sensor, one 16x2 LCD character display, one piece 10K potentiometer, two pieces 220 Ohms, and one piece 150 Ohms resistors.
Day 5	Four pieces 1.5V batteries, one 4 pieces battery holder, one push button, one small DC motor, one piece N2222 transistor, one piece IN4001/IN4007 diode, two pieces 150 Ohms resistors, one piece 10K Ohms resistor
Day 6	One sound detection sensor module, one 5V 10 amp AC relay, one AC light bulb (with holder), some insulated AC power wires
Day 7	One consumer remote control set (or a normal TV remote), one TSOP1738 (or equivalent) infrared receiver, one SM0038 IR receiver, one infrared transmitter LED (Blue/Transparent/other color), one NPN transistor (such as a N2222/BC547 general purpose NPN transistor)
Day 8	An additional Arduino Uno R3 board with USB A to USB B cable, two NRF24L01 2.4 GHz MSI band RF modules, one HC-04 Bluetooth module, three pieces 1K resistors
Day 9	One SIM800 GSM module (Arduino compatible), one active SIM card, one GSM module matching DC power source (Either battery or AC to DC adapter)
Day 10	One ESP8266 Wi-Fi chip, Wi-Fi router or/and Internet sharing capable smart phone

Table 1: List of hardware components

Things you will learn in this book

After completing this book, you will have learnt the following types of projects, devices, and fundamentals.

The following are the types of projects:

- Self-contained micro-controller projects
- Interfacing with single peripheral devices (such as sensors)
- Building compound devices (multiple devices in a single setup)
- Prototyping standalone devices (powered from independent power sources)
- Working with actuators (such as DC motors)
- Interfacing with AC powered devices (a light bulb)
- Using transmitters, receivers and transceivers
- Short range wireless communications (using Bluetooth and radio frequency)
- Long range wireless communications (using GSM modules)

The following are sensors/device examples:

- External LEDs
- Photodiodes
- Photo resistors (LDRs)
- DHT11 temperature and humidity sensor
- Soil moisture sensor
- MQ2 smoke detector
- Piezo Buzzer
- SD card module
- 16x2 LCD
- HC-SR05 ultrasonic sensor
- DC motor
- Push button
- AC-DC relay
- Sound detector
- TSOP1738/TSOP1838 IR receiver
- SM0038 IR receiver
- IR transmitter LED
- NRF24L01 RF module
- HC-04 Bluetooth module
- GSM SIM800 module
- ESP8266 Wi-Fi module

The following is the list of fundamentals/concepts:

- Arduino C programs/sketches
- Measuring voltage via analog pins from resistors, diodes, and transistors
- Using serial window for viewing program output and debugging
- Ohm's law calculations
- Bread boarding
- Voltage based logical HIGH/LOW
- Using homogeneous voltage devices that is, 5V devices with Arduino Uno R3
- Using device datasheets
- Working with Arduino libraries
- Digital input/output
- Analog input/output
- Using interrupts
- Reverse current in motors
- Voltage division technique
- Using external power supply
- Using multiple power sources
- Concept of pull up resistors
- Concept of pull down resistors
- Concept of **Pulse Width Modulation (PWM)** for motor speed control
- Using multiple power sources in a single setup
- Common grounding
- Keeping DC and AC powered circuits separate
- Working with **Infrared (IR)** signals
- Prototyping **Radio Frequency (RF)** devices
- Working with a Bluetooth device
- Wi-Fi module
- Basic AT commands
- GSM module
- Internet of Things
- IoT edge device prototyping
- IoT Cloud

Summary

In this first chapter, we have understood who should benefit from using this book. We have also seen how the chapters in this book are structured in a well-planned manner, so that the reader can gain maximum knowledge in a minimum time frame.

By the end of this book, our aim is to have made you proficient in prototyping with the Arduino platform in just 10 days, and we urge you to take on the challenge! After completing the fundamental project types in this book, you will be in a state to build almost anything on your own. Some of the interesting areas that you will be ready to build after mastering the topics in this book will be infrared and radio frequency remote controlled gadgets, smart retail projects, smart environment projects, and Internet of Things projects.

From the next chapter, we shall begin our journey of learning prototyping on the Arduino platform where we will be introduced to the basics.

2
The Arduino Platform

"Start from wherever you are and with whatever you have got." - Jim Rohn

The second chapter of this book is dedicated to laying down a strong technical foundation for the years to come in your Arduino based device development. To start off, we will see what the Arduino platform is all about. Then there will be quick introductory topics regarding fundamentals for getting started. This will be followed by a quick but in-depth look at the first Arduino program (known as a Sketch in the Arduino world).

Things you will learn in this chapter :

- Explanation of the Arduino platform
- Basic parts of an Arduino board
- Procedure to setup an Arduino board
- Writing a basic C sketch (program) for Arduino
- Blinking the onboard LED (via pin 13)
- Setting pin modes for peripheral devices
- Logical HIGH/LOW voltage values
- Arduino current supply limits
- Flash Memory (EEPROM) based program area
- Digital input and output techniques
- Analog input and output techniques

Introduction to the Arduino platform

The Arduino platform is a hardware board that offers a hardware-software integrated creative device development platform. The Arduino boards can be interfaced with openly available peripheral devices and custom programs can be written and loaded (embedded) into the Arduino development board:

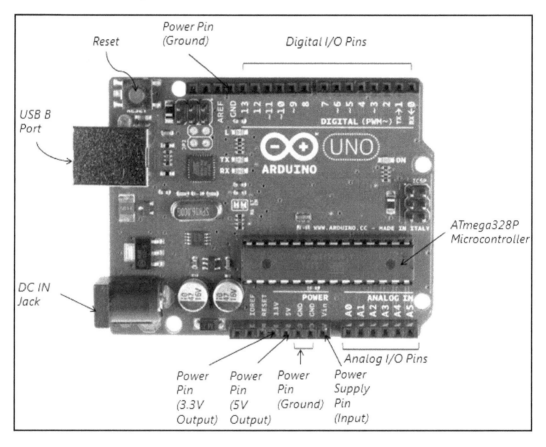

Figure 1: A typical Arduino Uno development board

These programs in turn control and interact with various peripheral components attached with the Arduino board. Thus by using the Arduino development boards; device designers and engineers can quickly create a running prototype of their imaginations. The best part is that the Arduino hardware and software eco-system is open source.

There are many different types of Arduino boards available in the market. Depending upon the processor speed, number of general purpose input/output pins and different power supply options there are many types of the Arduino boards. These boards are chosen depending on a project's needs.. The most commonly used and recommended Arduino board for the purpose of learning is the **Arduino Uno board**.

Usually Arduino Uno boards have an **ATmega328P** microcontroller and surrounding the microcontroller there are several **General Purpose Input Output** (**GPIO**) pins into which male jumper wires can be plugged in for connecting to peripheral devices.

The ATmega328P microcontroller is used as the brain of the Arduino Uno board. It is an 8 bit 5 volt **Reduced Instruction Set** (**RISC**) based single chip microcontroller. The ATmega328P microcontroller is also referred to as an AVR microcontroller which is a family of microcontrollers developed by **Atmel**. Atmel is a popular micro controller manufacturer.

Apart from the GPIO pins, you will notice several electronic components on the board such as a voltage regulator and some **Integrated Circuits** (**ICs**) to name a few. You will also notice a distinct USB B port and a DC IN power socket.

Let us get to know the Arduino Uno board and its components a little better, so that we are aware of what goes where during the course of this book. For a detailed list of components, the Arduino Foundation website is always recommended for further reading on latest updates.

 For a complete reference of all the components you can visit the official Arduino Foundation website at `https://www.arduino.cc/en/reference /board`.

A list of the major components has been provided in the next table for quick and easy reference. The following description should suffice for getting started on the Arduino platform and also for the scope of this book. Throughout your prototyping journey, you will find this reference very handy and useful.

Arduino component(s)	Component usage
ATmega328P microcontroller	The microcontroller executes the embedded programs.
Digital I/O pins	There are a total of 14 digital I/O pins numbered from 0 to 13. Out of the 14 pins six pins provide **Pulse Width Modulation** (**PWM**) output. These pins help in providing a connection with various peripheral components. While creating device prototypes we will frequently plug in male jumper cables into the female I/O pins of the Arduino board.
Analog I/O pins	There are six analog input pins, numbered from A0 to A5. These pins are also used for connecting various peripheral components.
Interrupt pins	Digital pins 2 and 3 provide a mechanism to receive interrupt signals from peripheral devices.
Power supply pin (input)	Vin, this pin can be used to receive power from external DC supply in the range of 5 - 12 volts.
Power pin (5V output)	5V, this pin provides an output of 5V regulated DC supply, for peripheral devices that run on 5 volts.
Power pin (3.3V output)	3.3V, this pin provides an output of 3.3V regulated DC supply, for powering peripheral devices that run on 3.3 volts.
Power pin (Ground)	GND, there are three ground pins on the board. These pins are used to plug in the common grounding jumper wires, in order to obtain a common ground reference for all the peripheral components connected to the Arduino board.
Flash memory	The ATmega328P microcontroller has 32 KB of flash memory. The embedded C programs are loaded into this memory area.
USB B port	This port is used to connect with and power up the Arduino board from a computer USB A port.
DC IN jack	This port is for receiving DC power from various DC power sources.

Table 1: Important Arduino Uno components

Overview of Arduino prototyping

Developing and prototyping with the Arduino platform is similar to working on any other embedded system. A general overview of the prototyping process is depicted in the following diagram:

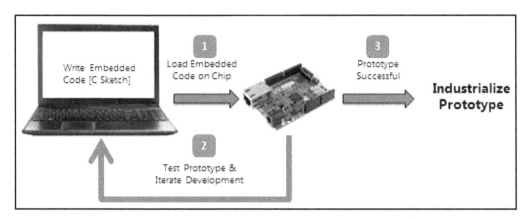

Figure 2: Overview of Device Prototyping with Arduino

The first step in the prototyping process is to write the embedded program that we will load into the micro-controller board. The term "embedded" is derived from something being hidden. Since the program is loaded into a micro-controller board which cannot be easily seen or deciphered, therefore the program is referred to as an embedded program. Once the program has been written and compiled successfully, it is loaded into the Arduino Uno board.

Once the program is compiled and loaded into the Arduino Uno development board, the next step is to setup the device prototype. A device prototype is the combination of the Arduino board together with other peripheral devices. The embedded program is written in such a way that it can control and interact with the peripheral components through the Arduino development board's **General Purpose Input Output** (**GPIO**) pins. After setting up the device prototype circuit, the next step is to power on the Arduino Uno board (if not done already). Once the board is powered on, the loaded program starts execution. In case of the Arduino a loaded program keeps running infinitely, as long as there is a power supply.

It is important to note that only one program (an Arduino program officially called a "Sketch") can be loaded into the Arduino Uno development board at a time. You cannot load more than one sketch at a time. Only one program keeps running infinitely. This is the typical way by which an Arduino sketch executes.

In case of the Arduino boards, simply connect the board with the computer USB port, using the USB A (on the computer) to USB B (on the Arduino) cable, as shown in the following figure:

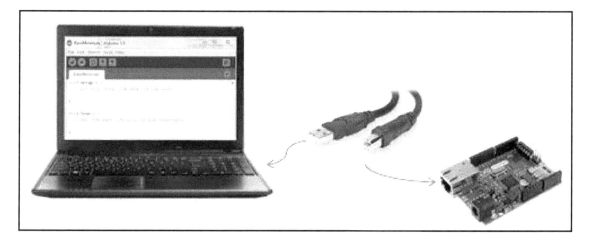

Figure 3: Connect Arduino with Computer via USB

Then launch the Arduino IDE on your computer. In the Arduino IDE, write the desired program (called a Sketch in the Arduino parlance). After the sketch has been written, compile and upload it into the Arduino board attached to the computer. While the sketch loads, the on-board LED on the Arduino board will continue flashing. Once the program has been loaded the flashing will cease automatically. From this point onwards, the loaded sketch will keep running infinitely, until there is a power supply.

In the following sections we will learn how to setup the Arduino board and how to start writing, compiling and loading the Arduino sketches into the Arduino development board.

Setting up the Arduino board

In this section we will briefly look at how to setup an Arduino Uno development board and its corresponding **Integrated Development Environment** (**IDE**).The process of the first time setup will depend upon the operating system of the computer and is available at the following Arduino Foundation website addresses. It is expected that the reader will perform the setup prior to proceeding further.

Depending upon the Operating System, the Arduino IDE can be downloaded from the following locations on the official Arduino Foundation website:

Windows: `https://www.arduino.cc/en/Guide/Windows`
Macintosh: `https://www.arduino.cc/en/Guide/MacOSX`
Linux: `https://www.arduino.cc/en/Guide/Linux`

As the majority of computer users worldwide use the Windows operating system, this book will provide some practical observations while setting up the Arduino Uno board on a Windows 7 computer (at the time of writing this book)--this information will help the reader additionally during the setup.

The following are the practical observations while setting up the Arduino Uno R3 board on a computer:

- After the installation is completed, and the board is connected for the first time, wait a bit for Windows to install the board drivers automatically. Give it about a minute. Don't rush.
- After the drivers get auto-installed by Windows, launch the Arduino IDE (you should find the shortcut on your desktop).
- Once the IDE is open, navigate to the **Tools** menu and perform the Board and Port selections as instructed in the setup steps. Usually, the board should get auto detected after plugging in the board to the computer. We will see how in the next section.

- However, sometimes the board might not get auto-detected, and this usually happens if the Arduino IDE is opened first followed by plugging in the board. However, there is nothing to worry about, just re-launch the IDE and the board should get auto-detected.

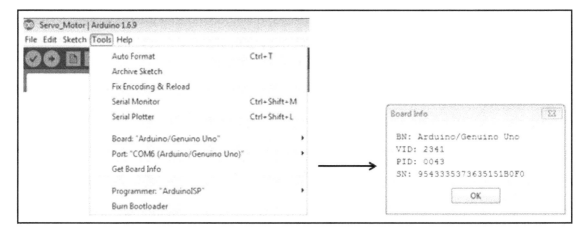

Figure 4: Tools | Get Board Info

- To verify, double check the board connection by navigating to the menu **Tools | Get Board Info**. If everything has been completed successfully, you should receive a pop-up, as shown in the preceding figure, with the board details such as Serial Number and so on.

Arduino program structure and execution

In this section we will study the general structure of an Arduino sketch. After going through this section you will be able to understand the important parts of an Arduino sketch. This section will also explain how each part of an Arduino sketch gets executed in the Arduino development board.

Arduino C programs are called **sketches**. A basic sketch structure needs at least two functions in the program body. These functions are listed as follows:

- `setup()`
- `loop()`

These functions should always have a same name that is, `setup()` and `loop()`. The Arduino development board is pre-programed to execute a loaded sketch by looking for these two function names in the body of a sketch. So if these two function names are not present in the C sketch, then the Arduino IDE will not even compile the sketch successfully. Instead the following errors will be thrown:

- undefined reference to `setup'
- undefined reference to `loop'

The next figure shows a standard Arduino IDE window and C sketch, on a Windows 7 operating system based computer.

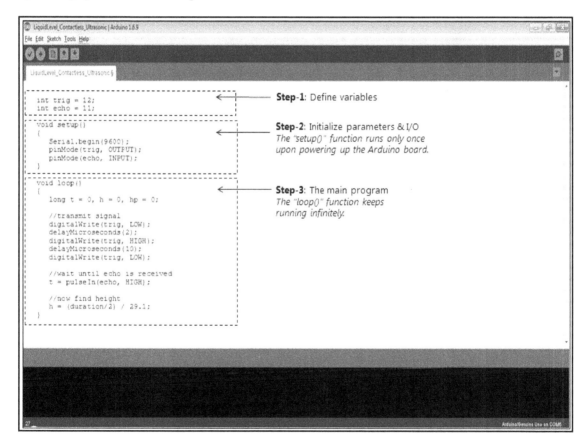

Figure 5: Arduino Sketch Basic Structure

You will notice that in the IDE there is a C sketch displayed on the screen. The distinct parts of the Arduino sketch have been highlighted with dotted lined rectangles. Usually, all Arduino sketches follow this same format, with at least these two basic functions: `setup()` and `loop()`.

Now let us try to understand the basic structure further. The `setup()` function is executed once and only once, every time when the Arduino board is powered up. This function is automatically invoked by the Arduino board, only for the first time when the board is powered up. This is an ideal place to write all the code for one-time configuration of variables and I/O pins used in the sketch. In all the example sketches in this book you will notice that various input/output pins are configured appropriately in this `setup()` function.

Next comes the `loop()` function that keeps getting invoked infinitely by the Arduino board. Whatever you write in this `loop()` function will keep getting executed infinitely, until the power supply to the Arduino board is turned off or drained out. This function contains the main logic of the embedded program. This main logic can in turn invoke various user defined functions.

In case you are curious, the code shown in the preceding sketch is for a water level measurement device using an Ultrasonic distance measurement sensor. The use of sensors in general has been explained in the `Chapter 4`, *Day 2 - Interfacing with Sensors* in this book. An ultrasonic sensor has been used in the `Chapter 6`, *Day 4 - Building a Standalone Device*. While a sound detection sensor has been used in the `Chapter 8`, *Day 6 - Using AC Powered Components*.

Understanding the first Arduino sketch

Now let us jump into some hands-on activity. In this section we will study an existing Arduino sketch that shipped with the Arduino IDE. The sketch we are going to study is the world famous "blink" program, which is like the Hello World of the embedded programming world. The Arduino IDE installation comes preloaded with a lot of example sketches. All the example sketches are available for use from the `File > Examples` menu.

Once you attain a minimum level of understanding, you will be able to start exploring more of these examples on your own. Let us inspect and understand the blink program. Fire up the Arduino IDE on your computer and navigate to the `File > Examples > 01.Basic > Blink` menu. The Blink program will open in a new Arduino IDE window.

The first point to note in the program is the one time configuration that has been done in the setup() function:

```
pinMode(13, OUTPUT);
```

The above line of code tells the Arduino board to configure digital pin number 13 in output mode. This is an important step, as it sets up digital pin 13 for output of signals from the microcontroller board.

 Note that when a digital pin is configured in output mode, the Arduino board is ready to transmit a digital voltage out via the digital pin. A digital signal is transmitted by sending an equivalent output voltage from a digital pin. This digital voltage in turn is received by the pin of a peripheral component that is attached to the digital pin that sends the digital signal.

In the blink program's case, the onboard LED is internally wired to Pin 13; hence we do not have to make any additional wiring or connections for this setup. Hence all we need to do is to compile and load the sketch.

 I/O pin setup:
When writing an Arduino sketch the first thing to do is to identify and appropriately setup the input/output modes of the pins being used in the device prototype. The Arduino pins connected to a peripheral device must be explicitly set in input/output mode.

The next important part of the sketch is the loop() function, which houses the instructions that are repeated over and over again. This function runs infinitely and is exactly how a micro-controller program operates. In the blink program's loop() function you will be introduced to two important in-built functions that are very commonly used in Arduino sketches. The first function is:

```
digitalWrite(13, HIGH);
```

This above function sends a HIGH digital signal on the pin number specified. In this case a HIGH signal is sent out via digital pin 13. In turn the HIGH signal is received by the on-board LED and it starts to blink as the HIGH signal passes through it.

 It is important to keep in mind that logical HIGH and LOW levels on a Pin are defined by the magnitude of the voltage on the pin. Typically, a HIGH state is identified by more than 3 volts and a LOW state is identified by less than 3 volts.

The second function used in the `loop()` function is:

```
delay(1000);
```

The delay function halts the program flow for the specified number of milliseconds. In this case the program execution halts for 1 second (or 1000 milliseconds). After waiting for 1 second the following line of code gets executed:

```
digitalWrite(13, LOW);
```

The above line of code results in a digital LOW signal being written on digital pin number 13, which in turn results in the onboard LED to stop glowing. Once again the program executes the following line of code:

```
delay(1000);
```

Thus the program execution waits for 1 second (or 1000 milliseconds). After waiting for 1 second the loop function is invoked all over again. Thus the sequence of code in the `loop()` function keeps switching the on-board LED ON and OFF continuously.

 It is worthwhile to note that Arduino Uno pins can supply a maximum of ~50 mA (some pins supply even less) of current (commonly referred to as "power" in online reference materials). Be careful when connecting Arduino pins to a component that cannot tolerate too much current and might burn out if too much current is provided--like an external LED. We shall see such an example in the hands on sketch that involves blinking an external LED in the Chapter 3, *Day 1 - Building a Simple Prototype*. Similarly, one should not connect a heavy duty component like a DC motor directly to an Arduino pin, as in this case the Arduino pin may get damaged. We will study this aspect in Chapter 7, *Day 5 - Using Actuators*.

Now that we have understood the onboard LED blinking program in detail, let us follow the steps below for loading the above program into the Arduino Uno board.

Compiling, loading and running a sketch

The first step in the process is to connect the Arduino Uno development board to the Computer using the USB-A to USB-B cable. The USB-A end of the cable will go into the computer's USB-A port. While the USB-B end of the cable will get plugged into the USB-B port on the Arduino board.

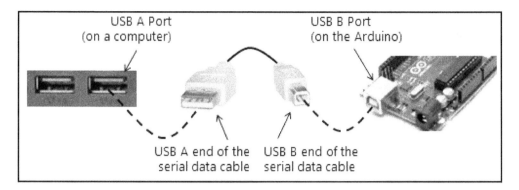

Figure 6: USB A to USB B Connector

After the Arduino board has been firmly connected to the computer, launch (or re-launch if already open) the Arduino IDE so that the Arduino board get auto detected by the IDE.

Navigate to the menu, **Tools | Port**, there should be an auto detected port, **COM6** (in the picture shown next). If the port is not selected by default, then go ahead and select it manually.

Sometimes it may take a few extra seconds for the board to get auto detected. If the board fails to get detected try to close the IDE and re-launch the IDE again.

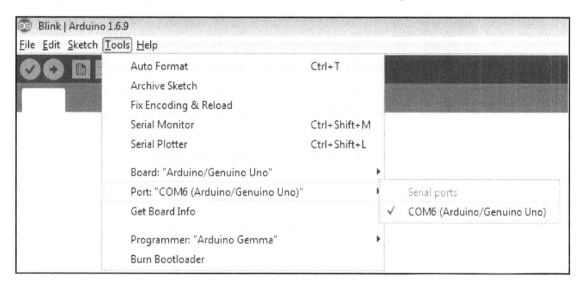

Figure 7: Selecting the Arduino Board

After the Arduino board has been detected and selected successfully (as shown above) the next step in the process is to compile the sketch. For compiling the sketch, click on the Tick Mark button, as shown in the following screenshot:

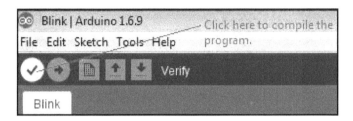

Figure 8: Compiling an Arduino sketch

Once the Tick Mark button has been clicked, the Arduino IDE will start compiling the sketch. In case the compiler finds any errors with the sketch it will provide a list of such errors in a standard error panel/window in the lower half of the IDE screen. You should look at the error list carefully and try to address the problems listed. Mostly the errors listed would be of a C programming nature:

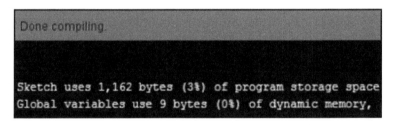

Figure 9: Upload to Arduino Board completed

Once the sketch has been compiled successfully, the Arduino IDE will display a message, as shown above, to indicate that the code compilation was successful. Upon successful compilation the Arduino IDE will also provide a message to display the amount of memory that would be consumed by the sketch.

 The important point to note here is that the Arduino Uno has a memory size of 32 KBs. The compiled program is loaded into this memory. Once more advanced scenarios are handled; the sketch has to be designed in a manner that it uses the 32 KBs of memory optimally.

After this the next step is to load the compiled code into the Arduino board. After compiling the sketch, the compiled sketch has to be uploaded in the Arduino board by clicking on the arrow button, as shown below. After the arrow button is clicked, the IDE will start uploading the C sketch into the Arduino board. Like any other embedded programmer, the IDE is actually loading the HEX code for the compiled C sketch into the board. HEX codes are the typical hexadecimal codes that are loaded into a micro-controller board for execution.

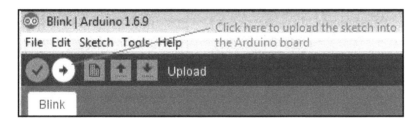

Figure 10: Upload a sketch to the Arduino board

Once the sketch is uploaded, the Arduino IDE will display a success message to indicate that the compiled code has been successfully uploaded into the Arduino board. Since the Arduino board is already connected to the computer via the USB port and is already powered up, the uploaded sketch starts executing right after getting uploaded.

Commonly used in-built C sketch functions

In this section we will quickly look at the commonly used in-built function in the Arduino C sketches. These functions will help us in performing basic tasks while interfacing with various devices. Unless you are attempting advanced sketches, the following functions will be very helpful in building most of the Arduino sketches. Apart from the below in-built functions, device specific header files and functions are also used - we will understand this aspect in the chapter for Day 2 while understanding how to use a header library for a sensor.

C Sketch function	Purpose
pinMode(PIN-NUMBER, I/O-MODE)	This function is used to specify whether a particular Arduino pin will be used in Input or Output mode.

C Sketch function	Purpose
`digitalRead(PIN-NUMBER)`	This function is used to read digital input from a digital pin. The input value is either LOW or HIGH and can be stored in a Boolean type variable.
`digitalWrite(PIN-NUMBER, SIGNAL-LEVEL)`	Use this function to send a digital LOW or HIGH signal on a particular pin.
`analogRead(PIN-NUMBER)`	This function is used to read analog input from analog pins. The input value can be stored in an integer type variable.
`analogWrite(PIN-NUMBER, SIGNAL-VALUE)`	Use this function to send an analog signal on a particular pin. The signal value can be specified as an integer between 0 and 1023.
`delay(MILLI-SECONDS)`	This function is used for halting the program execution for the specified number of milliseconds.
`delayMicroseconds(MICRO-SECONDS)`	This function can be used to halt the program execution for the specified number of microseconds.
`Serial.begin(BAUD-RATE)`	This function sets the baud rate for communication between the Arduino board and Arduino IDE's Serial Monitor window (also with any serial port communication program)
`Serial.println("MESSAGE")`	This method is used to output a string message to the Serial Monitor. It results in printing each message in a separate line.
`Serial.print("MESSAGE")`	This method is used to output a string message to the Serial Monitor. All messages sent are printed one after the other on the same line.
`isnan(VALUE)`	This in-built function is used to determine whether a specified value is a number or not.

C Sketch function	Purpose
`tone(PIN-NUMBER, FREQUENCY, MILLI-SECONDS)`	Use this function to play a sound of a particular frequency on a buzzer connected to a particular pin.
`pulseIn(PIN-NUMBER, LOGIC-LEVEL)`	Use this function to read and measure the duration of an input signal on a particular pin. Based on the logic level specified as a parameter to the function, it will wait for and read the signal until its logic level changes. For example, when reading a HIGH signal, it waits for a HIGH input signal and measures the duration until the signal level changes to LOW.
`millis()`	This function returns the number of milliseconds since the Arduino board was powered on and the program started running.

Table 2: Common in-built C functions

As we write more and more Arduino sketches you will start realizing that a large portion of the sketches deal with reading input and writing output to and from digital and analog pins. So it is important that we understand the basic techniques employed for the purpose of I/O. In the next two sections we will briefly explore these two options.

Digital input and output

In this section, let us look at the commonly used in-built C functions that are used for digital I/O. The following two functions are used for sending a digital signal out via a digital pin. First the digital pin has to be configured in output mode (for sending) in the `setup()` function:

```
pinMode(PIN-NUMBER, I/O-MODE)
```

Let us say we want to send a digital signal via digital Pin 7, then the following line of code will be used:

```
pinMode(7, OUTPUT)
```

This is how we configure the digital pin in output mode so that it can send digital signals. The next step is to understand how to actually send a digital signal. For sending a digital **HIGH** signal the following line of code must be used:

```
digitalWrite(7, HIGH)
```

Similarly, for sending a digital **LOW** signal the following line of code must be used:

```
digitalWrite(7, LOW)
```

Note that some peripheral components might need a timed signal pattern in order to get activated. For example, when using an **HC-SR04** Ultrasonic sensor, it needs a timed signal sequence in order to send a burst of ultrasonic waves. A timed I/O signaling scenario for the ultrasonic sensor is shown in the following figure:

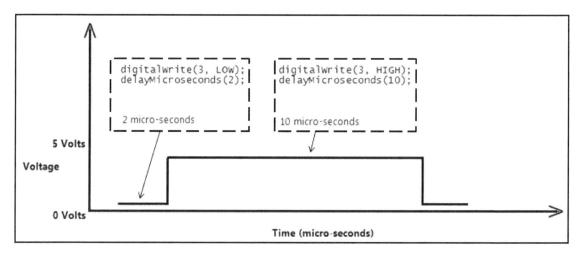

Figure 11: Timing sequence using Digital I/O

To be specific, the Ultrasonic sensor module needs a LOW signal for 2 micro-seconds, followed by a HIGH signal for at least 10 micro-seconds. This signaling pattern activates the Ultrasonic sensor. Using the digital signal functions we can easily achieve this by using the following piece of code:

```
// Signal a quick LOW just before giving a HIGH signal
digitalWrite(3, LOW);
delayMicroseconds(2);

// After 2 micro-seconds of LOW signal, give a HIGH signal
// to trigger the sensor
digitalWrite(3, HIGH);
// Keep the digital signal HIGH forat least 10 micro-seconds
```

```
// (required by HC-SR04 to activate emission of ultra-sonic
// waves)
delayMicroseconds(10);

// After 10 micro-seconds, send a LOW signal
digitalWrite(3, LOW);
```

Now that we are acquainted with the digital output technique, let us concentrate on the digital input techniques. The following two functions are used for receiving a digital signal via a digital pin. First the digital pin has to be configured in input mode (for receiving) in the setup() function:

```
pinMode(PIN-NUMBER, I/O-MODE)
```

Let us say we want to receive a digital signal via digital Pin 7, then the following line of code will be used:

```
pinMode(7, INPUT)
```

Now we have configured the digital pin in input mode so that it can receive digital signals. The next step is to understand how to actually receive a digital signal. For receiving a digital signal the following line of code must be used:

```
bool pinState;          // declare a variable
pinState = digitalRead(7); // read the state HIGH or LOW
```

The input value is either LOW or HIGH and can be stored in a Boolean type variable, as shown above.

Analog input and output

In this section, let us take a look at the commonly used in-built C functions that are used for analog I/O. To begin with we will learn how analog input is measured by the Arduino sketch. An Arduino sketch uses the following function to obtain an analog reading from its analog pins:

```
analogRead(<pin-number>)
```

For example, if we want to read the analog signal value from analog pin A5, we must use the following function call:

```
int value;              // declare integer variable
value = analogRead(A5);   // read analog value
```

The above function will return an integer value between 0 and 1023, depending upon the voltage value on the pin. The number 0 to 1023 will be proportional to the voltage range of the Arduino Uno board i.e. 0 to 5 volts.

Next we will look at techniques to write analog signals on both analog as well as digital pins. Digital pins that are PWM capable can simulate the behavior of an analog signal, hence the `analogWrite()` function can be used on a PWM capable digital pins as well. We shall understand the concept of PWM in depth in the chapter on Actuators during Day 5.

 For now, think of PWM as a technique to send analog signals of various strengths. Technically, the signal strength is referred to as the Duty Cycle of a signal. A 0% duty cycle means the signal is always OFF, a 50% duty cycle means the signal is ON half of the time and a 100% duty cycle means the signal is always ON.

For example, by using the following line of code we can send a full strength (100% duty cycle) signal.

```
analogWrite(5, 255);        // 100% duty cycle signal
```

Sending such a signal ensures that whatever component is being driven by the pin starts running at full speed (motors) or glows at full intensity (LEDs).

Whereas, by using the following line of code we can send a half strength (50% duty cycle) signal.

```
analogWrite(5, 127);        // 50% duty cycle signal
```

Sending such a signal ensures that whatever component is being driven by the pin starts running at half speed (motors) or glows at half intensity (LEDs).

While, by using the following line of code we can send a zero strength (0% duty cycle) signal.

```
analogWrite(5, 0);          // 0% duty cycle signal
```

Sending such a signal ensures that whatever component is being driven by the pin stops.

 The digital pins that are marked with the '~' sign are PWM capable. It is important to note that unlike digital output, the `pinMode()` function is not required in case of using the `analogWrite()` function. This is because the `analogWrite()` function automatically configures a pin to output mode before sending out the signal.

For sending an analog signal via analog pin A5, we will use the following line of code:

```
analogWrite(A5, 200)
```

Whereas, for sending an analog signal via digital pin 5, we will use the following line of code:

```
analogWrite(5, 200)
```

The signal value can be specified as an integer between 0 and 255. Based on the integer value the pin will send an equivalent PWM signal.

Try the following

Now that we have reached the end of this chapter, let us try to perform some interesting self-learning exercises. These exercises have been devised to simulate your thoughts and enable to utilize what you have learnt so far.

- Increase the delay between the blinking of the on-board LED to 2 seconds by using the statement delay(2000) in the `loop()` function.
- Decrease the delay between the blinking of the LED to 0.5 seconds. Remember 1 second = 1000 milliseconds.
- Wrap the LED blinking portion of the code in a `for` loop, to run 10 times.

Things to remember

A list of important concepts that you should remember has been provided so that you can easily recall the main points learnt in this chapter.

- In the Arduino parlance, a "program" is called a "sketch".
- Only one sketch can be loaded at a time into the Arduino board.
- The `setup()` function runs only once every time the board is either reset or powered up.
- The `loop()` function keeps running infinitely.
- Peripheral devices are connected using the digital and analog I/O pins.
- Peripheral devices may be powered by using the 5 volt and 3.3 volt power supply pins.
- The Arduino Uno board has 32 KB of memory. It must be used efficiently for managing complex scenarios.

- Arduino power supply pins should NOT be used for high powered devices such as motors.
- As the number of peripheral devices increase, the Arduino board's power supply pins may not be able to supply enough current (power). The chapter for Day-5 explains using batteries.
- When writing an Arduino sketch the first thing to do is to identify and appropriately setup the input/output modes of the pins being used in the device prototype.
- The `pinMode()` function is used to configure the pins for input and output.
- The `digitalRead()` and `analogRead()` functions are used for reading input signals from the pins.
- The `digitalWrite()` and `analogWrite()` functions are used for writing (sending) output from the pins.

Summary

In this chapter we concentrated on understanding the basics of writing a sketch; loading and executing the sketch in the Arduino board. We also learnt the most commonly used in-built C functions, followed by a detailed understanding to sending and receiving digital and analog signals. So far our journey was focused on the basics and the sketch that we studied was contained within the Arduino Uno board.

In the next chapter we are going to get introduced to building breadboard circuits with external hardware components and then learn how to interface them with the Arduino board. We will also take an in-depth look at various common electronic components used for Arduino prototyping. It is important to grasp the basic concepts as these will go a long way and make you fully self-reliant in your future endeavors.

3
Day 1 - Building a Simple Prototype

"A journey of thousand miles must begin with a single step."
-Lao-tsu

We will start this chapter by learning to build our first electronic circuit around the Arduino Uno board. This will be our first hardware-software integrated prototype. Two examples have been chosen for this chapter. The first example is a series of three external red **Light Emitting Diodes** (**LEDs**). We will write an Arduino sketch that will blink the three LEDs, one at a time, in a cyclic fashion. The second example is using a **Piezo Buzzer**. We will learn to write a sketch to play a musical tune using the Buzzer.

The reason for starting off with these basic examples is to understand how to build a prototype using the Arduino platform. And what better way could there be than to show some light (using the LEDs) and make some sound (using the Buzzer)!

After reading through this chapter, the reader will understand the fundamentals of using basic electronic components such as **resistors**, **transistors**, and **diodes**.

Things you will learn in this chapter:

- Use of resistors to reduce current flow
- Ohm's Law based resistance calculation
- Introduction to the breadboard
- Blinking multiple external LEDs
- Concept of common grounding
- Making sound with Buzzers

- Using transistors for switching
- Using rectifier diodes for preventing flow of current

The three LED project

If you are beginning your journey with micro controller-based prototyping, read on carefully. This discourse will empower you to jump start using the Arduino platform for rapid hardware software prototyping. It should be treated as a quick starter guide for using certain electronic components. However, for advanced electronics usage and calculations; an electronics engineering book must be referred to:

Figure 1: The three LED project

Before we start with the three LED project, the following section will provide a quick summary of why resistors will be used in this project.

Rationale for using a resistor

Resistors are commonly used to reduce the amount of current flowing from the source to the destination. To draw up an analogy, it is like squeezing a water supply tube at a particular point thus reducing the flow of water. Similarly, by placing a resistor across a piece of wire, the flow of current gets reduced.

Use a resistor to reduce the flow of the current in a circuit, in order to protect a delicate electronic component that cannot tolerate high currents.

The rule of thumb is to apply a resistor of an appropriate value, in series just before a delicate electronic component that cannot withstand too much current, for example an LED. We will look at an example of this fundamental concept in the following hands-on activity for blinking the three external LEDs in a cyclic fashion.

The two ends of a resistor do not have any polarity. A resistor can be placed in any direction in a circuit.

First things first, to begin with, we must understand why using a 220 Ohms resistor is necessary. Typically, a red LED is rated to work at 20 mA and has a voltage drop of 1.4 volts across it.

 Keep the classical **Ohm's Law** formula, shown in the following, in mind while trying to calculate the adequate value of the resistance to be employed.
Resistance = Voltage / Current

When receiving current from a 50mA source (Arduino pins in this case), the current needs to be reduced before it is allowed to flow through the LED. If the amount of current is not reduced then the LED will burn out eventually.

Ohms law is used, as described in the following, to determine the required value of the resistor:

Resistance = (Supply voltage - LED voltage drop) / Current

= (5 volts - 1.4 volts) / 0.020 amps

= 3.6 volts / 0.020 amps

= 180 Ohms

~ 220 Ohms (nearest standard resistor value)

 Although beyond the scope of this book, you will have to get used to determining the value of a resistor by looking at the color bands. In the beginning, when you work out of a starter kit, everything will be fine for a few weeks as all the resistors will come labelled properly.
However, after a few months, as you start building more prototypes, it is very likely that you will lose track of which one is which. In order to avoid any confusion, additional reading is suggested online, for decoding the resistor color bands by visual inspection.

For this project, the following parts will be required:

- Arduino Uno R3
- USB connector
- 1 breadboard
- 3 red LEDs
- 3 pieces 220 ohms resistors
- Some male-to-male jumper wires

Once all the required parts are available, go ahead and assemble them as shown in the following breadboard diagram:

In the diagram, you will notice the logo **fritzing** mentioned in the image. The fritzing software is an open source program used for creating circuit diagrams. While you are mastering the basics outlined in this book and starting your journey with Arduino, fritzing will help you to quickly draw your circuit diagrams.

Throughout this book, the basic breadboard diagrams have been created using fritzing. However, some additional labeling has been done outside fritzing. Overall, using fritzing is highly recommended for your device prototyping journey:

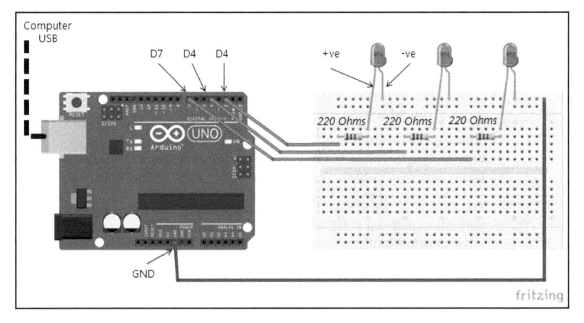

Figure 2: Wiring of the three LED project

The connection details between the Arduino Uno board and the three LEDs are provided in the following table for reference:

Arduino Uno pin	LED pin(s)
Digital pin 2	+ve (longer) leg of the 1st LED
Digital pin 4	+ve (longer) leg of the 2nd LED
Digital pin 7	+ve (longer) leg of the 3rd LED
GND	-ve (shorter) legs of the 3 LEDs

Table 1: Arduino to LED pin mapping

Let us understand the **breadboard** circuit connections in detail. When connecting LEDs, the thumb rule is to connect the longer leg of the LED to the positive terminal of the power source. The longer leg of an LED is the positive terminal of the LED. In our case, we are going to connect the positive terminals of the LEDs to digital pins of the Arduino board.

The digital pins of the Arduino board will supply the required current to make the LEDs glow. The shorter leg of an LED is the negative terminal. The negative terminal is connected to the ground. In our case, you will notice that the shorter legs of the three LEDs have been connected to the GND (ground) pin of the Arduino board.

 Note that all the shorter legs or the negative terminals of the three LEDs are connected to Arduino's GND (ground) pin via a common wire. This concept is referred to as **common grounding**. The basic idea is that all components in a circuit should have a common ground or reference for operating correctly. This is an important concept and will be very useful when you start building more complex prototypes with independent power sources. An advanced example of common grounding will be presented in the Chapter 7, *Day 5 - Using Actuators*.

After assembling the circuit, load the following sketch into the Arduino Uno board. It is advisable not to connect any wires to the Rx (pin D0) and Tx (pin D1) pins of the Arduino board while loading the sketch.

 Do not connect anything to pins 0 (Rx) and 1 (Tx) while loading the sketch. This is because these two pins are internally used as the hardware serial lines while the sketch gets loaded from the computer's USB port to Arduino's memory. For the scope of this book, this is the basic precaution that must be taken while loading sketches. Hardware serial communication is the transfer of data between the computer and the Arduino board via the USB ports. Serial communication has been explained in greater detail in Chapter 11, *Day 9 - Long Range Wireless Communications.*

Go ahead and load the following C sketch. This code may be freely downloaded from the GitHub location mentioned in the Chapter 1, *Boot Camp* of this book.

```
// Step-1: No variables are used in this sketch

//*********************************************************/
// Step-2: INITIALIZE I/O PARAMETERS
//*********************************************************/
void setup()
{
  // initialize digital pins as an output.
  pinMode(2, OUTPUT);
  pinMode(4, OUTPUT);
  pinMode(7, OUTPUT);

}
//*********************************************************/
// Step-3: MAIN PROGRAM
//*********************************************************/
// the loop function runs over and over again forever
void loop()
{
  digitalWrite(2, HIGH);   // turn LED 1 ON
  delay(1000);             // wait for a second
  digitalWrite(2, LOW);    // turn LED 1 OFF
  delay(1000);             // wait for a second

  digitalWrite(4, HIGH);   // turn LED 2 ON
  delay(1000);             // wait for a second
  digitalWrite(4, LOW);    // turn LED 2 OFF
  delay(1000);             // wait for a second
```

```
digitalWrite(7, HIGH);    // turn LED 3 ON
delay(1000);              // wait for a second
digitalWrite(7, LOW);     // turn LED 3 OFF
delay(1000);              // wait for a second
}
```

After loading the sketch successfully, re-connect the jumper wires to the Arduino pins; connect the GND pin in the end. As soon as you connect everything and power on, the three LEDs will start going on and off at 1 second intervals, one by one. As you gain more experience with loading Arduino sketches, you will realize that only the jumper wires to D0 and D1 need to remain unplugged while loading a sketch; and all other jumper wires can stay where they are. However, for the first few times, to be extra careful, unplug all the jumper wires when loading a sketch.

The preceding C sketch is a simple improvisation of the first blink program. In this sketch, three digital pins have been used and each digital pin has been made HIGH and LOW at intervals of 1 second--thus blinking the three LEDs one by one.

Let us understand the sketch in detail. The sketch has the typical three sections:

- A section to declare variable
- The setup() function
- The loop() function

However, in this case the starting section does not have any variables; therefore, this first portion of the sketch is empty. The setup() function configures the digital pins 2, 4, and 7 in output mode. So, the Arduino board is now ready to transmit output voltages via the digital pins 2, 4, and 7.

Note how the pinMode(pin-number, mode) function has been used to configure the digital pins of the Arduino board. For analog pins, the same pinMode(pin-number, mode) function can be used to configure the analog pins for the desired mode. The mode can either be INPUT or it can be OUTPUT. In INPUT mode, the Arduino pin will be ready to receive an input voltage from an external component.

The loop() function is fairly straightforward. To start with, it simply sends a HIGH signal on digital pin number 2, using the following function:

```
digitalWrite(2, HIGH);    // turn LED 1 ON
```

Then the sketch instructs the micro controller to halt for 1 second by using the following line of code. Hence the LED remains in a glowing state for 1 second:

```
delay(1000);                // wait for a second
```

Next, the sketch sends a LOW signal on digital pin number 2, using the following function:

```
digitalWrite(2, LOW);    // turn LED 1 OFF
```

Just like last time, once again the sketch instructs the micro controller to halt for 1 second by using the following line of code. Hence the LED remains in an OFF state for 1 second:

```
delay(1000);                // wait for a second
```

Similarly, the preceding sequence of instructions is repeated for the remaining two digital pins 4 and 7. Overall, the sketch switches the three LEDs to ON and OFF at 1 second intervals repeatedly. Thus, creating an effect where it seems the three LEDs keep glowing in a cyclic manner.

The example we just saw has numerous applications in the **electronic signage industry**. One must have seen commercial displays and billboards where an innovative sequence of glowing lights and LEDs capture the imagination of onlookers. The Arduino platform would come in as a very handy tool while giving shape to such a digital display project.

Another large area where the preceding example will help is when developing devices to indicate the status of various conditions. For example, think of a manufacturing machine in a mobile phone assembling factory. The manufacturing machines have many moving parts that are controlled via fluid mechanics. The fluid is supplied from a central tank. It is very important to maintain a sufficient amount of fluid in this central tank at all times. Think of a device that must be built to measure the fluid levels continuously. If the fluid level falls below an acceptable range then the device must activate a warning status by blinking a series of red LEDs on a warning panel attached to the manufacturing manager's office.

Just think about the limitless possibilities and you will be amazed to find how many commonly used household gadgets and electronic equipment have a blinking LED!

In the previous sections, we learnt how to control LEDs with the Arduino board, by making use of some in-built C functions. In the next section, we will look at another simple component, the Piezo Buzzer, for making sounds and playing tunes.

The Piezo Buzzer project

In this section, we will learn how to use a **Piezo Buzzer** and make sounds using some in-built C functions. A Piezo Buzzer is like a small speaker capable of emitting sounds at various frequencies. A typical Piezo Buzzer is shown in the following figure:

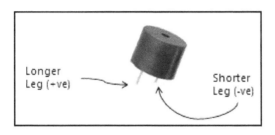

Figure 3: A Piezo Buzzer

A Piezo Buzzer has two legs, one longer positive leg and one shorter negative leg. The negative leg is connected to the ground while the longer positive leg is usually connected to an Arduino pin via a 100 Ohms resistor.

 Sometimes, you may find a small sticker pasted to the top of a brand new Piezo Buzzer that says, "Remove after washing". Do not get confused by this, it is just a leftover sticker from the Buzzer manufacturing process, which was not removed. Just ignore and remove the sticker. Do not wash the Buzzer!

For building the Piezo Buzzer project, we will use the following hardware components in this example:

- Arduino Uno R3
- USB connector
- 1 breadboard
- 1 Piezo Buzzer
- 1 piece 100 ohms resistors (higher resistor values will lower the sound intensity)
- Some male-to-male jumper wires

You may substitute the 100 ohms resistor with a higher value resistor. However, doing so will result in the reduction of sound intensity. Once all the required parts are available, go ahead and assemble them as shown in the following breadboard diagram:

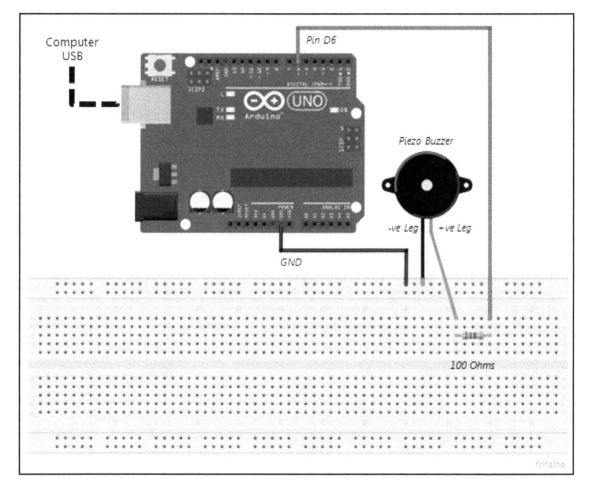

Figure 4: Wiring of the Buzzer project

Notice how the longer positive leg of the Piezo Buzzer has been connected to the digital pin number 6. This has been done so that the Buzzer can receive power from Arduino's pin number 6.

 It is worthwhile to note that Arduino Uno's I/O pins can safely provide upto 40 mA of current. Beyond this level, damage might start happening to the Arduino board. So be careful about what you connect to the pins.

While connecting a component like an LED, a Buzzer, or a small sensor directly to the Arduino I/O pins may be fine, one must avoid connecting heavier components directly. Heavier components are like motors that can draw more than 40 mA.

The connections between the Arduino and the Piezo Buzzer are listed in the following table for reference:

Arduino Uno pin	Buzzer pin(s)
Digital pin 6	+ve (longer) leg
GND	-ve (shorter) legs of the 3 LEDs

Table 2: Arduino to Buzzer pin mapping

While making the connections, you will notice that one leg of the LED is longer than the other. The longer leg of the LED is the positive terminal of the LED. The resistor should be placed between the Arduino pin and the positive terminal of the LED.

After assembling the circuit, load the following sketch into the Arduino Uno board. As a reminder, it is advisable not to connect any wires to the Rx and Tx pins of the Arduino while loading the sketch. In the next section, we will learn how to use in-built C functions to interface with the Piezo Buzzer:

```
//*********************************************************/
// Step-1: Variables used in this sketch
//*********************************************************/
//The following frequency values for musical notes have been
//collected from the Arduino Foundation website
//address mentioned below
//https://www.arduino.cc/en/Tutorial/ToneKeyboard?from=Tutorial.Tone3
#define NOTE_C4   262
#define NOTE_CS4  277
#define NOTE_D4   294
#define NOTE_DS4  311
#define NOTE_E4   330
#define NOTE_F4   349
#define NOTE_FS4  370
#define NOTE_G4   392
#define NOTE_GS4  415
#define NOTE_A4   440
#define NOTE_AS4  466
#define NOTE_B4   494

int pin = 6;          // the pin used to send signals to the buzzer
```

```
//***********************************************************/
// Step-2: INITIALIZE I/O PARAMETERS
//***********************************************************/
void setup()
{
  // no special setup related code in this example
}



//***********************************************************/
// Step-3: MAIN PROGRAM
//***********************************************************/
// the loop function runs over and over again forever
void loop()
{
  tone(pin, NOTE_C4, 500);
  delay(1000);
  tone(pin, NOTE_D4, 500);
  delay(1000);
  tone(pin, NOTE_E4, 500);
  delay(1000);
  tone(pin, NOTE_F4, 500);
  delay(1000);
  tone(pin, NOTE_G4, 500);
  delay(1000);
  tone(pin, NOTE_A4, 500);
  delay(1000);
  tone(pin, NOTE_B4, 500);
  delay(1000);
}
```

In a nutshell, the preceding C sketch plays the seven basic fundamental notes in a musical octave. This sketch utilizes the frequency definitions that are available on the official Arduino foundation website. Let us analyze the sketch and see how exactly the musical notes are getting played through the Buzzer.

To start with, the sketch starts with the top section where the variables used in the program are declared. For sounding the Piezo Buzzer, we have utilized the following function:

```
tone(pin, frequency, duration)
```

This function takes the `pin` number through which a sound with the specified `frequency` will be played on the Buzzer for the specified `duration`.

Therefore, it is very important that we know to what frequency the seven fundamental musical notes (that is, C, D, E, F, G, A, and B) corresponds to. This information can be readily found on the official Arduino foundation website. A reference URL is mentioned at the top of the sketch. You may visit this location and search for frequency data for more musical notes.

The `setup()` function in this case is empty, as we do not need to explicitly configure any pins. This is because the pin configuration is taken care of by the `tone()` function.

Once in the `loop()` function, the program executes the following line of code to sound a musical note on the Piezo Buzzer. The variable `NOTE_C4` corresponds to the frequency `262`, as defined at the top of the sketch. Hence, the following line of code results in a sound of frequency 262 Hertz being played for a duration of 500 milliseconds:

```
tone(pin, NOTE_C4, 500);
```

Thereafter, the program flow pauses for a second by using the following line of code. This pause has been given intentionally, in order to keep a small silence between two corresponding notes. It is up to you to increase or decrease the gap as well as duration of the sounds, depending upon how you want the sequence of notes to play out:

```
delay(1000);
```

The preceding two lines of code are repeated seven times, once for each musical note.

With this information in mind can you now imagine how we can build a simple electronic keyboard? You thought right! We can use seven buttons and wire them up with the Arduino, and on pressing each button, the sketch should sound a musical note of an appropriate frequency. Similarly, you may apply this knowledge in so many other projects that require sounds to be emitted. The possibilities are limitless.

In the following sections of this chapter, two other commonly used components have been described in a very clear manner. These components (transistors and diodes) are commonly used while building Arduino prototype circuits. It is very important that you grasp the fundamentals in this chapter, so that you can understand the rationale behind their usage in the remaining chapter in this book.

Using transistors

Transistors are commonly used for **switching**. They act just like a physical ON/OFF switch. Transistors are a separate subject altogether, hence the focus will be to stick to the basics, to the extent needed to start building prototypes with the Arduino.

A typical transistor has three legs, known as: Collector, Base, and Emitter. The most common types of transistors are classified as PNP and NPN transistors. If you hold the flat face of a transistor towards yourself, then for a PNP type transistor, the legs are arranged in the order--Collector, Base, and Emitter (from left to right); whereas for a NPN transistor, the legs are arranged as Emitter, Base, and Collector (from left to right).

A transistor is built in such a way that internally there is a circuit between the Emitter, Base, and Collector. Normally, a transistor's internal circuit is in an open (non-conductive state).

However, when a small voltage is applied to the Base, the internal circuit between the Emitter, Base, and Collector gets closed (conductive state)--hence switching ON the flow of current through it. Thus, a transistor acts as a switch and gets activated (closed conductive state) when a voltage is applied on the base.

 Applying a voltage on a transistor's base will close the internal circuit of a transistor, hence allowing flow of current. It is analogous to lifting/pressing a physical light-bulb switch to the ON position with your finger. In the context of the Arduino, you may send a small voltage to the base of a transistor by using the `digitalWrite()` or the `analogWrite()` function.

When using the `digitalWrite()` function, the Arduino pin sends a HIGH signal to the base of the transistor to which the digital pin is connected. Whereas, when using the `analogWrite()` function, the Arduino pin sends a modulated square wave signal (from a digital pin) or an analog signal (from an analog pin), to the base of the transistor to which the I/O pin may be connected.

Once a signal is received by the base leg of the transistor, the circuit between the Collector and the Emitter legs get closed. Thus, the transistor gets switched ON for the duration, until which the voltage remains applied on the base leg. As soon as the voltage is stopped on the base, the transistor gets switched OFF. In this section, we are going to perform a hands-on activity for understanding how transistors are used for switching.

The parts required for this activity are listed as follows:

- Arduino Uno board
- USB A to USB B cable
- 1 red LED
- 1 piece 220 Ohms resistor
- 1 N2222 transistor (other NPN transistors may be used as a substitute)

- 2 pieces 150 Ohms resistors
- Breadboard
- Some male-to-male jumper wires

After all the components have been assembled, the next step would be to follow the breadboard setup shown in the next diagram and build the prototype accordingly. Remember that the flat face of the transistor should face towards you, as shown in the following figure:

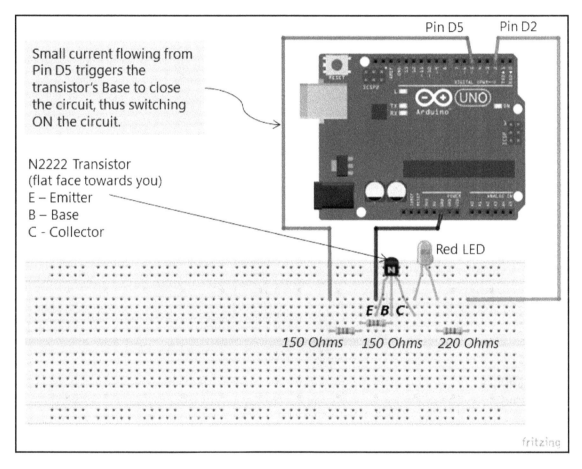

Figure 5: Using transistors

Notice how the transistor's base is connected to Arduino's digital pin number 5 via 300 ohms resistors. This is because we want a small voltage to be applied on the base of the transistor for the purpose of switching. We will see how in the following explanation.

After the breadboard circuit has been assembled and built as shown in the preceding figure, the next step would be to load the following sketch into the Arduino board. The following sketch can be downloaded from the GitHub location for this book:

```
//*******************************************************/
// Step-1: Variables used in this sketch
//*******************************************************/
int LEDPin = 2;          // specify the digital pin number
int transistorPin = 5; // specify pin for switching ON
// the transistor base

//*******************************************************/
// Step-2: INITIALIZE I/O PARAMETERS
//*******************************************************/
void setup()
{.
  pinMode(LEDPin, OUTPUT);        // Pin 2 as output
  pinMode(transistorPin, OUTPUT); // Pin 5 as output
}

//*******************************************************/
// Step-3: MAIN PROGRAM
//*******************************************************/
// the loop function runs over and over again forever
void loop()
{
  digitalWrite(LEDPin, HIGH);       // turn LED 1 ON
  analogWrite(transistorPin, 200); // turn transitor ON
  delay(1000);                      // wait for a second
  digitalWrite(LEDPin, LOW);      // turn LED 1 OFF
  analogWrite(transistorPin, 0);// turn transistor OFF
  delay(1000);                      // wait for a second
}
```

Program listing 2: Using transistors example

After loading the preceding sketch, you will notice that the LED starts blinking. Let us understand how. The following line of code is actually responsible for switching the transistor ON, so that the LED can glow:

```
analogWrite(transistorPin, 200); // turn transitor ON
```

The preceding line of code sends a steady square wave signal via the PWM capable digital pin number 6. Upon receiving the signal from pin 6, the transistor base gets closed and completes the circuit between the Collector and the Emitter. Thus, the transistor gets switched ON.

Similarly, the following line of code switches OFF the transistor, as the voltage on digital pin number 6 is diminished to zero:

```
analogWrite(transistorPin, 0);   // turn transistor OFF
```

Since there is no longer a voltage on the transistor base, the circuit between the Collector and the Emitter becomes open. Thus, the transistor gets switched OFF.

In order to practically see how the `analogWrite()` function plays a role in switching the transistor ON and OFF, just comment out the following line of code in the preceding sketch by adding the double front slash symbols (//) at the beginning of the line, as follows:

```
//analogWrite(transistorPin, 200); // turn transitor ON
```

Once the preceding line of code has been commented, load the sketch again into the Arduino board. This time you will notice that the LED does not glow anymore. This is because a voltage is no longer getting applied on the base of the transistor from Arduino's pin number 6.

With this, we will conclude our study of using a transistor for switching in Arduino-based prototypes. What we learnt in this section is a globally used technique for switching in electronic circuits. You may apply the same fundamentals of using a transistor for switching with other micro-controllers as well.

In the real world, multiple devices can be independently switched ON and OFF using this technique. More transistor examples will be provided in `Chapter 7`, *Day 5 - Using Actuators*.

Using diodes

Semiconductor diodes are of many types and are a separate topic beyond the scope of this book, hence we will proceed with the basic understanding and build upon it as we proceed. We will focus on **rectifier diodes** for the scope of this book.

 The primary function of a rectifier diode is to act just like a check valve allowing current to flow in one direction and resisting (up to a limit) current from flowing in from the opposite direction.

In practical situations, sometimes under certain situations a reverse current is generated--such as when a motor armature is stopped, it continues to spin (due to inertia) inside the motor case until it comes to a halt.

This reverse current has the potential to flow back into other electronic components. This reverse current might lead to damage of delicate electronic parts. In order to stop this reverse current from flowing back, a diode is placed in an appropriate manner in the circuit. Let us try to understand what a diode does exactly by going through the following hands-on exercise.

We will require the following parts for conducting this exercise:

- Arduino Uno board
- USB A to USB B cable
- 1 red LED
- 1 pc. 220 ohms resistor
- 1 pc. IN4007 diode (other IN400x version diodes may also be used)
- Breadboard
- Some male-to-male jumper wires

After all the components have been assembled, the next step would be to follow the breadboard setup shown next and build the two prototypes, one by one. There are actually two separate breadboard setups as shown in the following diagram:

The circuit on the left-hand side demonstrates how a rectifier diode can be placed to allow current to flow, while the circuit on the right-hand side demonstrates how to use a rectifier diode to block the flow of current.

Notice how the banded or striped edge of the rectifier diode has been placed in the two circuits. The placement of this banded or striped edge makes all the difference in either allowing or not allowing the incoming current to flow through:

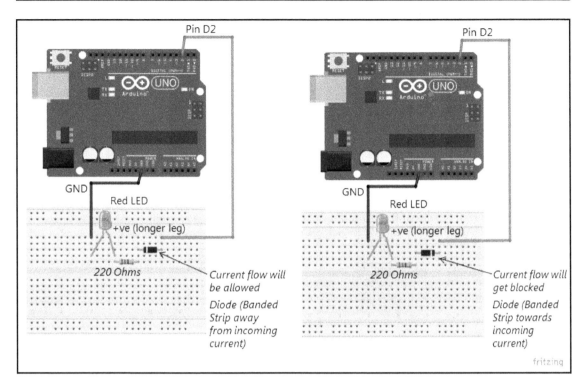

Figure 6: Using rectifier diodes

Let us begin with the circuit on the left-hand side. After assembling the circuit on the left-hand side, load the following sketch into the Arduino board. Remember not to keep any jumper wires connected to the Rx and Tx pins while loading a sketch into the Arduino board:

```
//************************************************************/
// Step-1: Variables used in this sketch
//************************************************************/
int LEDPin = 2;          // specify the digital pin number

//************************************************************/
// Step-2: INITIALIZE I/O PARAMETERS
//************************************************************/
void setup()
{
  // initialize digital pins as an output.
  pinMode(LEDPin, OUTPUT);
}

//************************************************************/
```

```
// Step-3: MAIN PROGRAM
//******************************************************/
// the loop function runs over and over again forever
void loop()
{
  digitalWrite(LEDPin, HIGH);    // turn LED 1 ON
  delay(1000);                   // wait for a second
  digitalWrite(LEDPin, LOW);     // turn LED 1 OFF
  delay(1000);                   // wait for a second
}
```

After the program is loaded and starts executing, you will notice that the LED starts blinking. This is because in the circuit setup, the banded or striped edge of the rectifier diode has been placed away from the positive voltage terminal (which is the digital pin number 2 in this setup). The current is allowed to flow through the diode in this case.

Now let us inspect the circuit that has been shown on the right-hand side. All you need to do in this circuit setup is to reverse the direction of placing the diode. In the right-hand side circuit setup, the rectifier diode has been placed in such a way that it will not allow current to flow through it.

In this setup, the banded or striped edge of the rectifier diode faces the incoming current from digital pin number 2. The banded or striped edge acts as a check valve, it does not allow current to enter. Thus, in this circuit setup, you will notice that the LED does not blink any longer. Just try to quickly reverse the diode the other way around and see how the LED starts blinking again. With this example, we conclude our quick study of using rectifier diodes.

In the previous sections, we covered two very important electronic components that are very commonly used in Arduino device prototyping circuits. The use of transistors and diodes will be further demonstrated for controlling a DC motor from an external power source, later in the book, specifically in Chapter 7, *Day 5 - Using Actuators* where we will learn how to use actuators.

In the last section of this chapter, we will see how to use simple push buttons in our circuits with the Arduino board. The next example that we will learn is to blink an LED when a button is pushed.

LED with a push button

It is very common to find buttons on every device all around us, be it a simple kitchen timer or a complicated microwave oven. Buttons provide users the ability to push them in order to instruct the device to do certain operations.

There are many types of buttons available on the market depending upon the usage scenarios. We will use a simple push button for the purpose of learning the fundamentals of wiring a button to a micro-controller and then receiving its input and thereafter triggering a piece of code to make a physical device do something. A typical push button is shown in the following figure for reference.

As shown in the next figure, a simple push button can be easily connected to the Arduino board by following the markings shown in the previous diagram. The important thing to note here is that, the upper leg is connected to a digital pin on the Arduino board. This digital pin will be used to read the signal level (high or low) available from the button leg. So, the digital pin will receive a HIGH signal as soon as the round button is pushed. Until the button is pressed, the signal will be LOW on the digital pin:

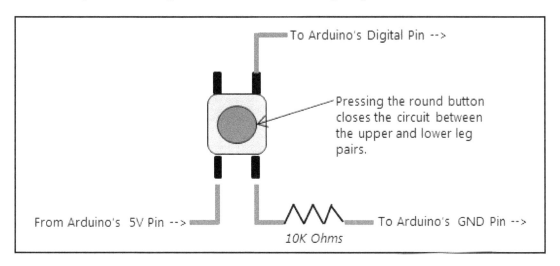

Figure 7: A typical push button

Utilizing the preceding information, let us quickly assemble a prototype to blink an LED as soon as a push button is pressed. You can refer to the breadboard setup shown in the following for reference.

The LED connection details have already been explained previously in this chapter, you may reuse it from the previous sections. The connection details of the push button to the Arduino board are provided in the following table for reference:

Push button legs	Arduino Uno pins	Comments
Top right leg	Digital pin D7	Connect directly
Top left leg	N/A	Not Used
Bottom left leg	5V	Connect directly
Bottom right leg	GND	Connect via 10K ohms resistor

Table 3: Push button to Arduino Uno connections

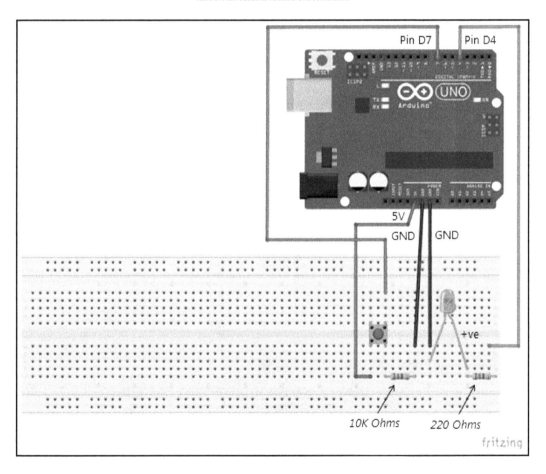

Figure 8: Button operated LED

Now let us look at the sketch that will read the input signal from the button and send a signal to the LED to glow for a second:

```
//*********************************************************/
// Step-1: CONFIGURE VARIABLES
//*********************************************************/
int buttonPin = 7;          // D7 used to read input from button
int ledPin = 4;             // D4 used to control LED
int buttonState = LOW;      // Initialize state to LOW to start with


//*********************************************************/
// Step-2: INITIALIZE I/O PARAMETERS
//*********************************************************/
void setup()
{
  Serial.begin(9600);
  pinMode(buttonPin, INPUT);  // D7 will receive input from button
  pinMode(ledPin, OUTPUT);    // D4 will send output to LED
}


//*********************************************************/
// Step-3: MAIN PROGRAM
//*********************************************************/
void loop()
{
  buttonState = digitalRead(buttonPin);  // Read button state

  // If user presses the push button
  // Then state will be HIGH
  if(buttonState == HIGH)
  {
    digitalWrite(ledPin, HIGH);  // glow the LED
    delay(250);                  // wait for 1 second
    digitalWrite(ledPin, LOW);   // switch OFF the glowing LED
  }
  delay(1000);
}
```

The first step in the preceding sketch is to define the two pins that have been used in the setup. Therefore, two local variables have been defined for storing the pin numbers to which the button and the LED have been connected.

The next step in the sketch is to configure the pins used in the correct mode. Since D7 will be used to read the input from the button, this pin has been configured in the input mode. Pin D4 will be used to send signals to the positive terminal of the LED, therefore it has been configured in the output mode.

The pins used for input or output must be identified and configured accordingly in the `setup()` function.

Thereafter in the `loop()` function, the program keeps looping until it finds the button to be pressed using the following line of code:

```
buttonState = digitalRead(buttonPin);   // Read button state
```

Once a HIGH signal is detected, it means that the button has been pressed. Hence the next part of the sketch is to blink the LED in the portion of the sketch using the following `if` statement:

```
if(buttonState == HIGH)
{
  digitalWrite(ledPin, HIGH);   // glow the LED
  delay(1000);                  // wait for 1 second
  digitalWrite(ledPin, LOW);    // switch OFF the glowing LED
}
```

Thus, we have a simple working prototype whereby pressing a button, we can make LEDs glow. With this example, we come to the end of this chapter. Going forward you may apply the same method to sound a Buzzer or pretty much activate any device on the press of a button!

Try the following

Let us try some exciting exercises to further enhance our understanding of the concepts that we learnt in this chapter:

- Try to increase an additional LED and modify the relevant portions of the sketch
- Try to decrease an LED and modify the relevant portions of the sketch
- Find out the maximum number of LEDs that may be blinked using an Arduino Uno

Try to use all the digital pins and modify the relevant portions of the sketch.

Try to play a famous song line like *Happy Birthday to you* using the Piezo Buzzer. First find the musical notes (for example, GGAGCBGGAGDC) of the song. Then use the corresponding frequencies for the notes. So, the note "G" would mean the variable NOTE_G4 used in the Buzzer example earlier.

Things to remember

Remember the following important points. These points will help you to quickly recall what we learnt in this chapter:

- Use a resistor to reduce current flow
- Ohm's law is very important to remember so that you can calculate the value of the required resistance in a circuit
- All peripherals in a circuit are connected to a common ground
- Transistors are used like electronic switches, operated by applying a voltage to their base
- Rectifier diodes are used as check valves, in order to allow current to flow only in one direction, while preventing current from flowing in the reverse direction.
- The banded or striped edge of a rectifier diode acts like a check valve to stop current from flowing in.

Summary

In this chapter, we concentrated on understanding the basics of building simple prototypes with the Arduino Uno board. We looked at how common electronic components are used with detailed hands-on examples. We learnt how to calculate the value of a resistor in a circuit, followed by how to use a transistor for switching and a rectifier diode for preventing current flow.

We also looked at two basic examples that included working with LEDs, Piezo Buzzers, and push buttons. Equipped with this knowledge, you will feel more confident about dealing with various Arduino-based breadboard circuits and their corresponding sketches.

In the next chapter, we are going to get introduced to interfacing our Arduino board with sensor modules in general. We will look at the fundamental process of interfacing with a sensor module by using its datasheet and Arduino library.

4

Day 2 - Interfacing with Sensors

"Equipped with his five senses, man explores the universe around him and calls the adventure science.

- Edwin Powell Hubble

A sensor is a special kind of electronic device that senses external stimuli such as heat, humidity, moisture, and light. Sensors are one of the integral building blocks of smart devices and embedded systems. One might have seen automatic doors that slide open once somebody goes near the door. These automatic systems are usually based on sensors, microcontrollers, and embedded software. In this chapter, you will learn the fundamental technique of interfacing and the use of sensors in general, with the Arduino platform. After mastering the basic steps, you will be able to work with more sensors on your own.

The following topics are covered in this chapter:

- Introduction to sensors
- Using elementary sensors
- Using integrated sensor modules
- General steps for interfacing sensor devices
- Using a datasheet for a sensor device
- Installing a library for using a sensor device
- Using the Serial Monitor to display sensor output
- Working with a temperature-humidity sensor
- Using a soil moisture sensor
- Using 5V devices with Uno R3

Types of sensor components

We will look at two very different types of sensors in this chapter, which are as follows:

- Basic sensor components
- Integrated sensor modules

In the following sections, we will explore these basic types of sensors and some useful techniques for interfacing with such sensor devices.

Basic sensor components

A **basic sensor component** can be thought of as an elementary electronic component that is used to measure electrical properties generated as a result of external stimuli on the component. An apt example of such a sensor would be electronic components such as photodiodes and thermistors. When external stimuli, such as light, heat, and moisture, act upon an elementary electronic component, then certain electrical responses are generated depending on the component and the stimuli. For example, temperature affects the resistance property of a thermistor. Similarly, when light is an incident upon a photodiode, it triggers the flow of current. In the first part of this chapter, we will learn how to measure the response generated in a photodiode and a photoresistor when light falls upon it.

Using a basic sensor - photodiode

Let us start our journey of learning basic sensor devices using a **photodiode LED**. A photodiode LED is usually black in color and looks slightly larger than a normal LED. A photodiode LED generates small amounts of voltage when light is incident upon it. The amount of voltage generated is proportional to the amount of light incident upon the photodiode. You will learn how to do an Arduino sketch to measure the voltage generated on the analog pin that is connected to the photodiode.

For easy reference, the parts required for this project are listed as follows:

- Arduino Uno
- USB cable
- Photodiode
- Ohms resistor
- Breadboard
- Some male-to-male jumper wires

Once all the preceding parts are assembled, build the circuit shown in the following breadboard figure:

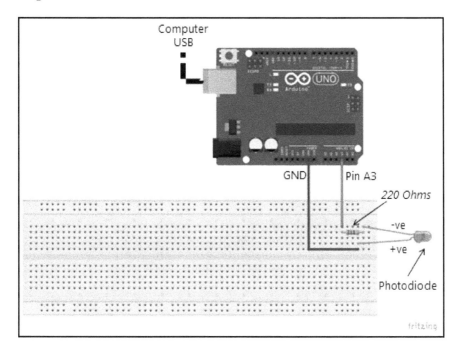

Figure 1: Using a photodiode with Arduino

The connection between the photodiode and the Arduino pins is provided in the following table:

Photodiode	Arduino Uno pin	Connection remarks
Positive leg	GND	Connect directly to the GND pin
Negative leg	Analog Pin A3	Connect via 220 Ohms resistor

Table 1: Photodiode to Arduino Uno connections

A few things must be noted in the preceding setup. The positive terminal (longer leg) of the photodiode is connected to the ground pin of the Arduino board, whereas the negative terminal (shorter leg) is connected to the analog pin number 3. This is exactly opposite of what we performed with a normal LED. This method of reversing the LED terminals is known as a reverse bias setup.

The important thing to understand here is the behavior of the photodiode. The photodiode will generate a measurable voltage as soon as light is detected. We will use this property of the photodiode to detect the presence and intensity of ambient light. Go ahead and load the following photodiode light detector sketch into your Arduino board:

```
//**********************************************************/
// Step-1: CONFIGURE VARIABLES
//**********************************************************/
int photoDiodePin = A3;        // analog pin 3
                               // to measure voltage from
                               // photodiode


//**********************************************************/
// Step-2: INITIALIZE I/O PARAMETERS
//**********************************************************/
void setup(void)
{
  Serial.begin(9600);         // begin serial connection
                              // so that out of this sketch
                              // can be received in Arduino's
                              // Serial Monitor
}


//**********************************************************/
// Step-3: MAIN PROGRAM
//**********************************************************/
void loop(void)
{
  // read the value at analog pin 3 to which the

  // photodiode is connected
  int value = analogRead(photoDiodePin);
  // analogRead() function returns a number which is proportional to
  // Arduino Uno's voltage which is between 0 - 5 volts
  // For example, 5.0 volts would return 1023
  // similarly 2.5 volts would return 511
  // Therefore the following calculation is used to convert the value
  // read from pin A3
  float voltage = (value / 1024.0) * 5.0;

  // The following statements is used to write text output to
  // Arduino Uno's Serial Monitor window
  Serial.print("Sampled Voltage: ");
  Serial.println(voltage);

  // Wait for 2 seconds before sampling the value on pin A3
  delay(2000);
}
```

After loading the preceding program, you will have to use the Arduino IDE's serial monitor window in order to display the output of the sketch. The USB port is the serial port, and the Arduino IDE's serial monitor window acts as a serial terminal client exchanging messages to and from the serial port, that is, the USB port leading to the Arduino board.

The following steps are outlined in order to launch the serial monitor window of the Arduino IDE:

1. Launch the Arduino IDE.
2. Navigate to **Tools |Serial Monitor**.
3. (Alternatively, you may simply click on the magnifying glass icon in the top right-hand corner of the IDE).

As soon as the Serial Monitor window is launched, the following outputs will start getting displayed on the Serial Monitor window, at every 2-second intervals. Now, let us understand how the sketch reads and calculates the voltage generated by the photodiode:

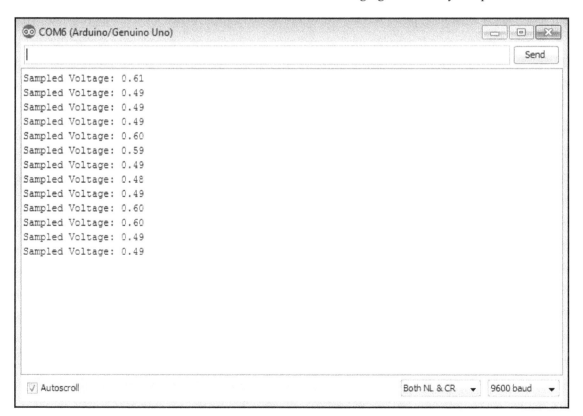

Figure 2: The photodiode readings

In the readings displayed in the preceding window, the values 0.59 and above were generated when a DC battery-operated light source (a simple handheld torch) was made directly incident upon the photodiode. Notice how the measured voltage spiked to 0.59 (and above) from the normal 0.49. The reading of 0.49 was obtained when the handheld light source was not incident upon the photodiode. So, here we see a direct relationship between the voltage generated and the intensity of incident light.

In the preceding sketch, the following line of code is responsible for reading the value at analog pin number 3, to which the photodiode is connected:

```
int value = analogRead(photoDiodePin);
```

The `analogRead()` function returns a number between 0 and 1023, which is proportional to Arduino Uno's voltage reference range, which is between 0-5 volts. For example, if 5.0 volts is available on an analog pin, then the number 1023 would be returned by the `analogRead()` function. Similarly, if 2.5 volts is available on an analog pin, then ideally the number 511 should be returned by the `analogRead()` function. While in case where 0 volts is available on an analog pin, then the number 0 should be returned by the `analogRead()` function.

The next important thing to do is calibrate the sketch to recognize and respond to a particular type of event; by "event", we mean the amount of incident light. The following are some examples that you may try:

- Ambient light at night
- Ambient daylight (during fully sunny weather)
- Ambient daylight (during fully cloudy weather)
- Shining a handheld torch light on the photodiode

In each case, you will notice that the measured voltage will be slightly different. The more the intensity of light, the higher would be the value of the voltage generated by the photodiode. In order to react to certain values, we can introduce an `if` condition to the `loop()` function in the sketch appropriately, as shown in the following code:

```
void loop(void)
{
  // read the value at analog pin 3
  // to which the photodiode is connected
  int value = analogRead(photoDiodePin);
  float voltage = (value / 1024.0) * 5.0;
  Serial.print("Sampled Voltage: ");
  Serial.println(voltage);
  // introduce special logic to handle lots of light
  if (voltage > 0.58)
  {
    Serial.println("There seems to be a lot of light! ");
  }

  // introduce special logic to handle lesslight
  if (voltage < 0.51)
  {
    Serial.println("There seems to be lesslight.");
  }

  delay(2000);
}
```

This calculation and understanding of how the voltage generated by incident light upon a photodiode can be utilized for measuring the output from other sensor devices or components as well.

Using a basic sensor - photo resistor (LDR)

Let us quickly look at one more example of using a photo resistor. Looking at this example will help us understand the use of elementary sensors in a better way. A **photo resistor** is also known as a **Light Dependent Resistor** (**LDR**). The resistance of an LDR varies with incident light. The resistance of an LDR decreases with an increase in the intensity of light. So, if the surroundings have very bright light, then the resistance of the LDR will be at its lower end. Whereas, when the surroundings are dark, then the resistance of the LDR will be at its maximum. Keeping this property of LDRs in mind, we can think of a way to measure the corresponding voltage as a result of the changing resistance, using any of Arduino's analog pins.

In the breadboard diagram shown next, notice how the photo resistor is connected to the Arduino board. A photo resistor has two legs of similar length. It does not have a positive or a negative terminal. Any of the legs can be connected to the 5V pin of Arduino, while the other leg will have two different connections: the first connection is to an analog pin, while the second connection is to the GND pin (via a 10K ohms resistor). Placing the resistor in this manner between the photo resistor's leg and GND will help us measure the voltage via the analog pin.

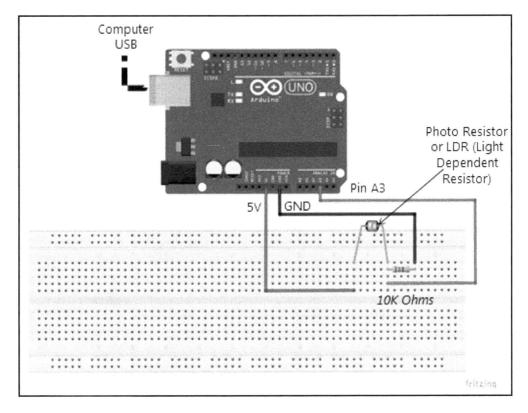

Figure 3: Using an LDR with Arduino

The connections between the photo resistor and the Arduino board pins are provided in the next table:

Photo resistor	Arduino Uno pin	Connection remarks
1st leg (right-hand side leg in the diagram)	GNG	Connect via 10K Ohms resistor
Analog Pin A3	Connect directly	Connects directly
2nd leg (left-hand side leg in diagram)	5V	Connect directly

Table 2: Photo resistor to Arduino Uno connections

After all the components of the preceding circuit have been assembled, load the Photo Resistor (LDR) sketch and use Arduino's Serial Monitor to view the output:

```
//***********************************************************/
// Step-1: CONFIGURE VARIABLES
//***********************************************************/
int LDRPin = A3;              // analog pin 3
                             // to measure voltage from
                             // photo resistor (LDR)


//***********************************************************/
// Step-2: INITIALIZE I/O PARAMETERS
//***********************************************************/
void setup(void)
{
  Serial.begin(9600);       // begin serial connection
                            // so that out of this sketch
                            // can be received in Arduino's
                            // Serial Monitor
}


//***********************************************************/
// Step-3: MAIN PROGRAM
//***********************************************************/
void loop(void)
{
  // read the value at analog pin 3 to which the photodiode is
  // connected
  int value = analogRead(LDRPin);
  // analogRead() function returns a number which is proportional to
  // Arduino Uno's voltage which is between 0 - 5 volts
  // For example, 5.0 volts would return 1023
  // similarly 2.5 volts would return 511
  // Therefore the following calculation is used to convert the value
  // read from pin A3
  float voltage = (value / 1024.0) * 5.0;
```

```
    // The following statements is used to write text output to
    // Arduino Uno's Serial Monitor window
    Serial.print("Sampled Voltage: ");
    Serial.println(voltage);

    // Wait for 2 seconds before sampling the value on pin A3
    delay(2000);
}
```

A sample Serial Monitor window with readings from the photo resistor is shown in the following screenshot. Notice how the values drop below 0.10 as soon as you cover the photo resistor surface with your hands:

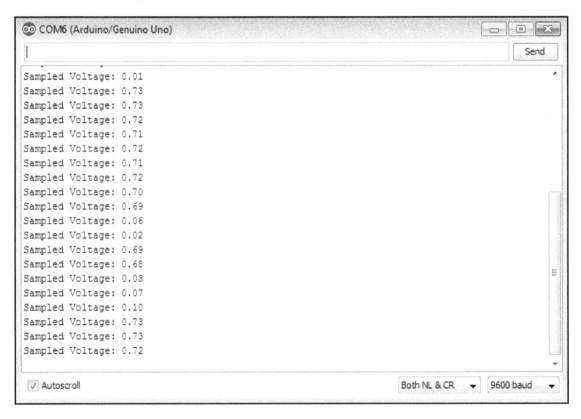

Figure 4: The photo resistor readings

Follow the instructions for launching Arduino's Serial Monitor and view the output to see how the sampled voltage varies with a change in the intensity of light.

As a practical example, you may create a device that will automatically switch on the lights in your backyard as evening arrives and the sun slowly sets. This can be done using a simple **Light Dependent Resistor** (**LDR**)and a Relay module with Arduino. We will see how to interface Arduino in the `Chapter 8`, *Day 6 - Using AC Powered Components*.

Similarly, there are elementary components such as thermistors, where the resistance reduces with heat. Thermistors can be used to measure temperature. Once again, the help of an external circuit is taken in order to measure the voltage generated by the changing resistance of thermistors.

Since Arduino's analog pins can be used to measure only voltage, some innovative circuitry is created around an elementary sensor component so that the corresponding voltage can be measured using Arduino's analog pins.

Depending upon the value measured by the analog pins and based on prior knowledge, study, and careful calibration, various physical phenomena (such as heat, light, and moisture)can be measured easily.

How the actual circuits are set-up with an elementary component is an advanced topic and would be more aptly placed in an electrical engineering book. Usually, sensor component manufacturers provide a technical specifications document, known as a **datasheet**. A datasheet provides various specifications of the components. In the case of integrated sensor modules, the datasheet usually contains a circuit diagram of how to connect the components with a microcontroller.

The information contained in a datasheet should be referred to while setting up the component's connections with the Arduino Uno board. While it will take some independent research for a beginner to get used to using basic/elementary components such as sensors, it is a lot easier to use the integrated sensor modules by referring to their datasheets.

In this half of the chapter, we learned the basic principle of using basic sensor components by measuring voltage changes via Arduino's analog pins. In the next half of this chapter, you will learn how to use an integrated sensor module by leveraging its datasheet and a corresponding Arduino library.

Using integrated sensor modules

An Integrated sensor module usually contains one or more elementary electronic components and additional hardware nicely designed and placed on a PCB unit. These chip-based sensor devices usually take care of the basic measurement and calculations required to arrive at a final measurement. When using these sensor devices, all one has to do is utilize a corresponding Arduino library that has been specifically written for the chip-based sensor device. In turn, the library hides all implementation details of the actual method of measuring the sensor values. We will look at one such example in this book. You will learn how to use a common temperature sensor module, known as **DHT11**. Additionally, in the last part of the chapter, we will work with another type of integrated sensor that does not require a specific Arduino library.

For using integrated sensor modules, in general, the following sequence of basic steps should be followed:

1. **Step 1:** Download the datasheet of the integrated sensor module.
2. **Step 2**: Determine the connection method of the integrated sensor module with a microcontroller (Arduino Uno, in this case).
3. **Step 3**: Install sensor, specific Arduino library.
4. **Step 4**: Write and load the Arduino sketch using the sensor's Arduino library functions.
5. **Step 5**: Connect the sensor to the Arduino board, as per the datasheet.

To start, one must refer to the datasheet of the integrated sensor module. As explained earlier in this chapter, the datasheet contains all the technical specifications of the integrated sensor module. A special area of interest in our case would be to understand how to connect the sensor to the Arduino board. As we proceed through this chapter, we will see how we can use the information in a datasheet.

Using a temperature sensor module (with an Arduino library)

Let us begin this lesson by taking the example of the **DHT11** temperature sensor module and inspect the steps required to use a sensor with the Arduino platform in general.

For quick and easy reference, the parts required for this project are listed as follows:

- Arduino Uno R3
- USB connector
- One breadboard
- One DHT11 temperature sensor
- One 5K Ohms resistor
- Three male-to-male jumper wires
- Three female-to-male jumper wires

Understanding sensor module datasheets

The datasheet is a fundamental piece of documentation that describes an electronic component in great detail. These documents are usually created and published by the device manufacturers and are freely available online. You may search for a specific device datasheet online and download the document.

Typically, a sensor device datasheet includes detailed descriptions, including a circuit diagram describing how to interface the sensor with a microcontroller. The datasheet is also very useful in finding out the operating conditions of a device.

Taking the example of the DHT11 sensor datasheet, it can be found by simply searching the phrase **DHT11 datasheet** in a search engine of your choice and then downloading the PDF file.

It is common to find this datasheet on several websites. The datasheet demonstrated in this book is from Micropik.com. At the time of writing this book the datasheet was available at the following link:

http://www.micropik.com/PDF/dht11.pdf.

 Before connecting an external device to Arduino, always refer to its datasheet. The datasheet for an electronic component is the detailed specifications document for that electronic component. It contains vital operational details such as current consumption and operating voltage.

Determining how to connect the sensor

The next step is to study the datasheet carefully and locate the page in the datasheet PDF file, where there is either a circuit diagram or an explanation of how to connect the sensor chip to a microcontroller. Once the connection details are found, it becomes fairly easy to connect the sensor with the Arduino board.

In the case of DHT11's datasheet, the following connection diagram can be found. Let us look at this connection diagram and understand how to use the information for connecting the DHT11 temperature sensor to Arduino Uno (the MCU or microcontroller, in our case):

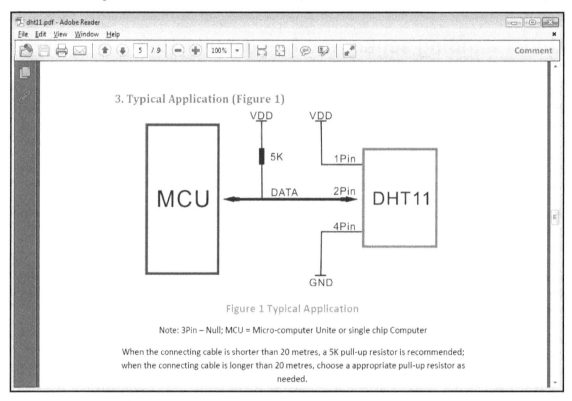

Figure 5: The DHT11 datasheet connection diagram

As shown in the preceding figure, the DHT11 sensor has four pins. It is important to notice that pin number 3 is missing in the preceding diagram and that there is a line that says - **Note: 3Pin - Null**. This means that the sensor module's pin number 3 is not required to be used for the connection.

After reading through the beginning of the datasheet, it will become clear that the DHT11 sensor module gives digital output. Hence, the output (pin 2) of the sensor should be connected to any digital pin of the Arduino board.

Based on the preceding diagram, the connection of each DHT11 pin to Arduino Uno is mapped in the following table:

DHT11 pin	Arduino Uno pin	Connection remarks
Pin 1	5V (VDD indicates the 5V pin on Arduino)	Connect directly to the 5V pin.
Pin 2	Any digital pin use digital pin 2 in this example (DATA indicates signals between the sensor and microcontroller)	Connect directly to digital pin 2. Also, place a 5K resistor between DHT11 pin 2 and Arduino's 5V pin.
Pin 3	N/A	This pin is marked as null which means that pin number 3 is unused and therefore it will not be connected to anything.
Pin 4	GND	Connect directly to any GND pin.

Table 3: Sensor to Arduino pin mapping

Connecting the 5K resistor adds a slight twist to setting up the DHT11 sensor with Arduino. Note how the 5K resistor connects pin1 and pin 2. The practical implementation of the preceding circuit diagram with Arduino is shown in the following diagram.

At this point, you must try to look hard and close at the connection between the schematic diagram in the datasheet and the breadboard circuit setup shown next and try to internalize how the indicative circuit diagram in the datasheet was translated into a practical setup shown using the breadboard:

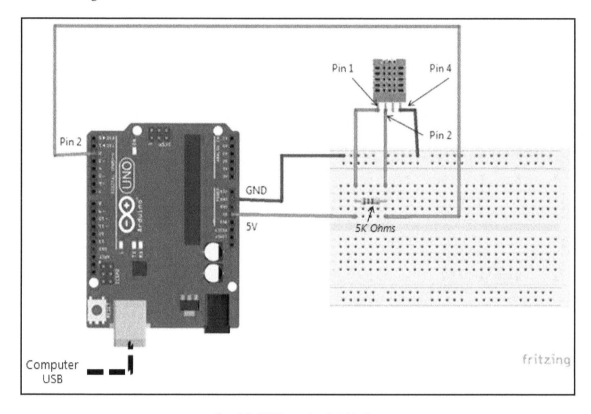

Figure 6: The DHT11 connection with Arduino Uno

Notice how the 5K Ohms resistor is connected between pin 1 and pin 2. Now, look back at the datasheet diagram and try to understand why this was done.

 It is very important that you do not rush through this phase. Stay here as long as it takes to grasp the fundamentals of connecting the sensor pins to the Arduino Uno pins, based on the sensor datasheet diagram.

After you have understood the circuit, the next step would be to look at the sensor-specific Arduino library. We will use the Arduino library for the sensor to read the temperature data from the sensor.

DHT11's datasheet will reveal that it is rated to work with a voltage supply of 3.3-5 volts. Devices that can tolerate incoming signals of 5 volts can be directly connected to the Arduino Uno pin unless the device has the potential of drawing a current beyond what can be supplied by the Arduino Uno pins (typically ~50 mA). Therefore, it can be observed that all the DHT11 pins were connected directly to the Arduino pins, without any stepping down of voltages.

Installing the sensor-specific Arduino library

After the hardware connections have been understood, the next step is to start developing the software part. When using sensors, usually most sensor devices need an additional piece of driver software in order to read the signals from the sensor device.

In the Arduino parlance, these software drivers are usually available as Arduino libraries. Most sensors (barring some) have their software drivers built and published online.

These Arduino libraries are available for free download via the Arduino IDE. In this section, the general approach for finding and installing an appropriate Arduino library for the DHT11 temperature sensor will be discussed. These steps can be reused for other sensors as well.

Install the DHT11 library for using the DHT11 temperature sensor with Arduino UNO. This library is required in order to include the DHT.h file in the Arduino sketch described in the next section.

In general, the following steps should be followed in order to use a header file in an Arduino sketch:

1. Launch the Arduino IDE.
2. Navigate to **Sketch** | **Include Library** | **Manage Libraries....**
3. The **Library Manager** window will pop up.
4. In the **Library Manager** window, search for the keyword DHT in the textbox auto-populated with the phrase, Filter your search....
5. Once the library is listed, go ahead and install it.

After installation, the library should appear as **INSTALLED** in this list, as shown in the following screenshot:

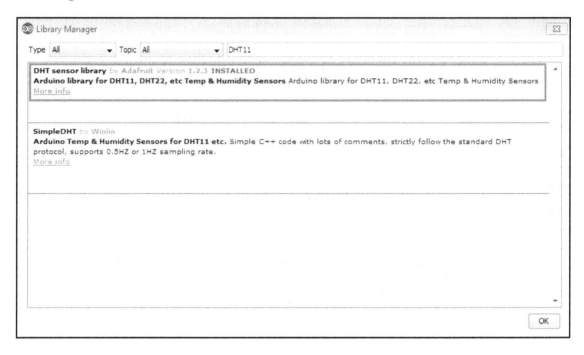

Figure 7: The DHT11 Arduino library

The preceding library has been used while writing the code in this book. Similarly, for other sensors, you should use the method explained previously for utilizing the sensor-specific header files.

Sensor interfacing sketch

This is the final phase of building the prototype. Let us inspect the following sketch for reading the values from the DHT11 temperature sensor. In this section, the Serial Monitor window of the Arduino IDE will be introduced. Understanding the Serial Monitor window is one of the fundamental steps in debugging the Arduino sketches.

Go ahead and load the following C sketch. As a best practice, avoid connecting anything to the Arduino Uno pins while loading a sketch.

The following DHT11 Sensor Sketch may be freely downloaded from the online location of this chapter mentioned in the Chapter 1, *Boot Camp* of this book:

```
//**********************************************************/
// Step-1: CONFIGURE VARIABLES
//**********************************************************/
#include "DHT.h"

#define DHTPIN 2
#define DHTTYPE DHT11

DHT dht(DHTPIN, DHTTYPE);  // create the sensor object


//**********************************************************/
// Step-2: INITIALIZE I/O PARAMETERS
//**********************************************************/
void setup()
{
  Serial.begin(9600);      // start software serial
                           // communication
  while (!Serial) { }
  Serial.println("Starting Temperature Monitor...");
  dht.begin();
}


//**********************************************************/
// Step-3: MAIN PROGRAM
//**********************************************************/
void loop()
{
  delay(5000);

  float t = dht.readTemperature();   // read temperature

  // check if temperature was read properly
  if (isnan(t))
  {
    Serial.println("Reading from DHT sensor failed");
    return;
  }

  // print the temperature on the Serial Monitor window
  Serial.print("Temperature: ");
  Serial.print(t);
  Serial.println(" *C");
}
```

After loading the sketch successfully, power off the Arduino board by unplugging the USB cable from the computer. Then, connect the DHT11 sensor pins using the jumper wires to the Arduino pins, as per the circuit diagram described earlier.

After connecting the circuit, simply power on the Arduino board by plugging in the USB cable to the computer. At this point, the Arduino sketch will start running.

Viewing the program output

The sketch execution will keep looping in the `setup()` function's line:

```
while (!Serial) { }
```

The preceding line of code instructs the Arduino microcontroller to check whether a serial connection is open. The C sketch keeps looping in the while statement until the Serial Monitor window is launched from the Arduino IDE.

As soon as the Serial Monitor is launched, the C sketch will stop looping in the while loop and move on to print the line, "Starting Temperature Monitor..." This line will be visible immediately on the Serial Monitor window.

After this, the `dht` object is started by the `dht.begin()` line of code. Thereafter, the `loop()` method keeps executing infinitely and samples the reading from the sensor at every 5-second interval.

If the datasheet is studied closely, then we will see that there is a section that states not to send any signals to the sensor within 1 second of powering up the unit as the system is in an unstable state during the first 1 second; to be cautious, one should double this time and wait for at least 2 seconds before querying the sensor - this requirement is covered by the `delay(5000)` statement.

Eventually, the `dht.readTemperature()` line of code reads the data from the sensor, and later in the sketch, this data is printed on the Serial Monitor window at every 5-second interval, as shown in the following screenshot:

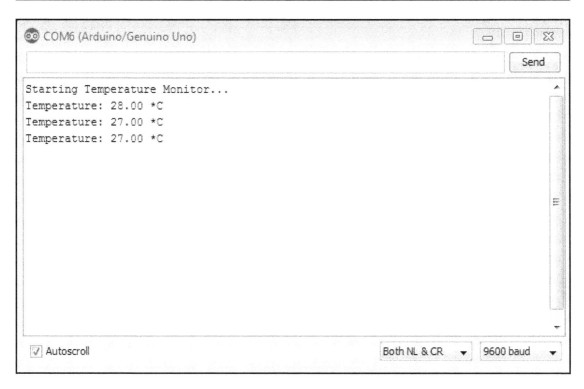

Figure 8: Displaying DHT11 sensor reading in the Serial Monitor window

 Sometimes due to loose connections on the breadboard, the readings might fail. However, if that happens, reconnect the components firmly and try again.

If the readings do not come as expected, then make sure that the jumper wires are firmly connected in place.

Using the Serial Monitor is a very convenient way of debugging Arduino sketches. Whenever you want to see the values of variables, you may utilize the `Serial.print()` or `Serial.println()` functions to output the values on the Serial Monitor screen.

In the foregoing example, we looked at the entire process of using an integrated sensor module in detail. Of special interest was the area where we learned how to use the sensor module's datasheet and the Arduino specific library.

However, there might be certain integrated sensor modules that may not require an Arduino specific library. Can you guess when? Well, the answer is usually when there is no special need to initialize the sensor device in a complex manner and/or when reading the output of the sensor module is fairly straightforward. One such example is the soil moisture sensor module. Let us understand how it works and how to write an Arduino sketch to use the sensor.

Using a soil moisture sensor module (without an Arduino library)

Now, let us quickly see an example of a sensor module that does not require an Arduino library. The sketch can read values from the sensor directly by simply utilizing built-in functions.

The following parts will be required to follow this example:

- Arduino Uno R3
- USB connector
- One breadboard
- One soil moisture sensor unit
- Some jumper wires

A typical soil moisture sensor unit has two separate parts, as shown in the following figure. One part is the moisture probe that will remain dipped in the soil. The probe is designed to measure soil moisture content using some advanced soil properties (such as a dielectric constant) of the soil.

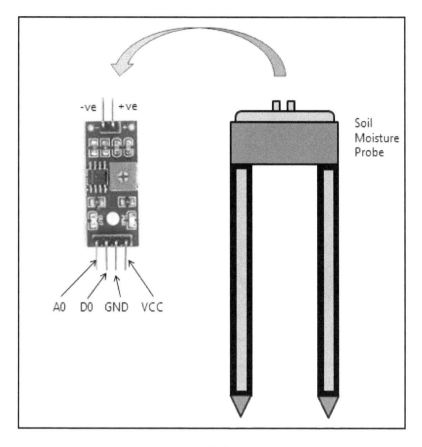

Figure 9: Parts of a typical soil moisture sensor unit

The second part of the soil moisture sensor is the integrated circuit that processes the electrical property variations from the probe. The measured electrical properties can be in turn read via Arduino's pins. The integrated circuit then provides four pins to be connected to the Arduino board. The pins are usually labeled very clearly. The A0 pin should be connected to an Analog pin, whereas, the D0 pin should be read through a digital pin on the Arduino board.

Soil moisture sensor circuit

Go ahead and assemble the breadboard circuit as shown in the following figure:

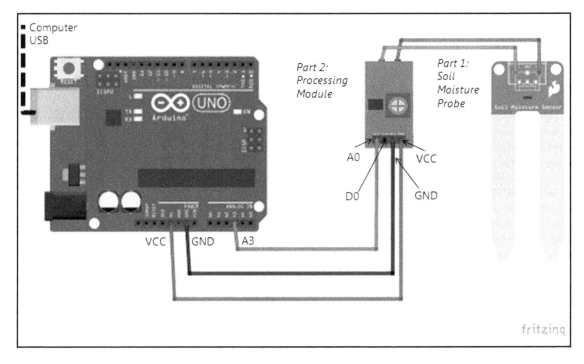

Figure 10: The soil moisture sensor circuit

A soil moisture sensor probe usually has two pins. However, some soil moisture probes may come with the second part integrated with the probe, in which case all the pins (even that of the second part) are available from the probe part. However, the basic connection between the soil moisture sensor and the Arduino board will fundamentally remain the same.

The connection between the probe and the second part is straightforward. The probe does not have a polarity marked on its pins. Hence, simply connect the two pins (any to any) with the two pins of the second part.

The connection between the second part and the Arduino board is provided in the following table for reference:

Soil Moisture pin	Arduino Uno pin	Connection remarks
VCC	5V	Connect directly to the 5V pin.
GND	GND	Connect directly to any GND pin.
D0	Not used	Not used in this example. However, it can be used via a digital pin.
A0	A3	Any analog pin may be used.

Table 4: Soil moisture to Arduino Uno connections

Now that we have understood how to connect the soil moisture sensor to the Arduino board, let us try to recall the fundamentals we learned in the previous chapters for writing the sketch to read the sensor.

Soil moisture sensor sketch

Can you think of the fundamental operation that we are going to do in this sketch? You may have already guessed, since we are using the soil moisture sensor's A0 that will be connected to Arduino's Analog pin; therefore, we will use the following built-in C function for writing the Arduino sketch:

```
analogRead(moisturePin);
```

The C sketch is provided for reference. You may download the following soil moisture sensor sketch shown below directly from the online location mentioned in the Chapter 1, *Boot Camp:*

```
//*****************************************************/
// Step-1: CONFIGURE VARIABLES
//*****************************************************/
// Analog Pin A3 will be connected to the A0 pin
// from the soil moisture sensor
int moisturePin = A3;

//*****************************************************/
// Step-2: INITIALIZE I/O PARAMETERS
//*****************************************************/
void setup()
```

```
{
  // Configure the A3 pin in input mode
  pinMode(moisturePin, INPUT);

  // Begin the serial connection so that
  // we can see the sensor measurements
  // on the Serial Monitor
  Serial.begin(9600);
}

//**********************************************************/
// Step-3: MAIN PROGRAM
//**********************************************************/
void loop()
{
  // Read the value from the A3 pin
  int sensorValue = analogRead(moisturePin);

  // Display the value on the Serial Monitor
  Serial.print("Soil Moisture Reading: ");
  Serial.println(sensorValue);

  // Wait for 5 seconds and repeat the steps again
  delay(5000);
}
```

You will notice that now the preceding sketch seems very simple. It is reusing some basic fundamental principles, such as:

- Configuring the appropriate pins in either an the INPUT or OUTPUT mode in the `setup()` function
- Reading voltage changes in sensor devices via Arduino's analog pin using the `analogRead()` function

Now, let us proceed further by loading the sketch into the Arduino board. Place the soil moisture sensor probe into a pot of dry soil. Alternately, you may also fill a cup with dry paper napkins and place the soil moisture sensor wrapped in the paper napkins inside the cup. Make sure the probe does not get fully submerged; submerging it to about 80% of its height should work fine. For the soil moisture sensor model used in this book, take caution not to drown the output pins from the probe.

For practical usage on the field, there are some soil moisture sensors available that may be submerged under water for a certain period of time - one such sensor is the **VH400** soil moisture sensor. So, if you are building a practical sensor for outdoor use, you may use this type of sensor. The sensor module used in this book should not be used in water-submerged conditions.

After everything is in place and the system starts running, open Arduino IDE's Serial Monitor for viewing the output from the sensor. A sample output on the Serial Monitor screen is shown in the following figure:

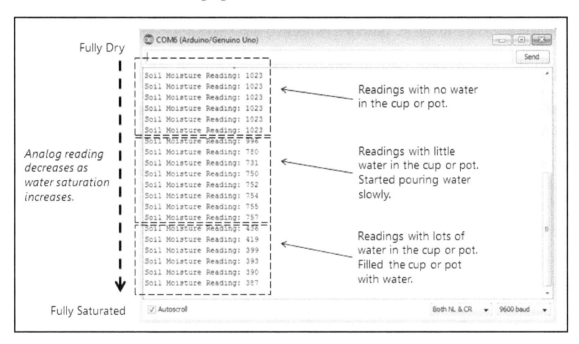

Figure 11: The soil moisture sensor readings

With this example, we will conclude our study of using various types of sensor devices with Arduino. If you follow these basic series of steps, then you can use other sensors and integrated devices with the Arduino platform in a similar manner.

Future inspiration

A real life example of using the soil moisture sensor would be in the area of smart irrigation systems. Smart irrigation systems are becoming very lucrative because they help in conserving water. Basically, the water outlets that feed irrigation canals are controlled by central microcontroller-based devices. These microcontrollers get soil moisture data from various sampling points spread over the agricultural lands. Depending on the saturation of water in the sampled area of land, the water outlets feeding the corresponding irrigation canals are opened. Thereafter, the water outlets remain open until the soil moisture saturation level reaches an acceptable level. The water outlets are usually opened and closed using solenoid valves controlled via AC powered relay devices. Later in this book, you will learn how to use AC powered devices via relay modules.

Now, to make the irrigation system smart, the microcontroller can check the weather channel forecast by consuming the weather channel's web service over wireless internet methods. If there is a forecast of rain in the next 12 hours, then the water outlets are not opened, thus saving water! Later in this book, you will learn about wireless communications and connecting to the internet. Thus, the microcontroller unit orchestrates the entire activity of the smart irrigation system.

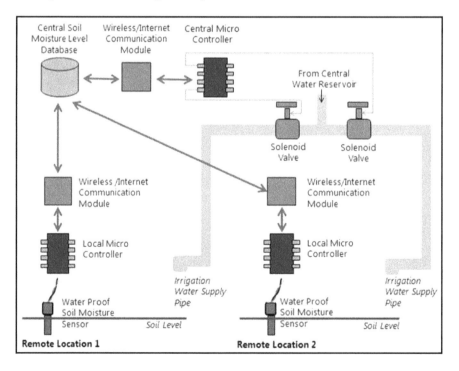

Figure 12: The smart irrigation system

After completing this book, you will be able to use the Arduino platform as a very handy tool to rapidly develop a working prototype of the preceding real-world example of a smart irrigation system.

This chapter showed us how to measure soil moisture. `Chapter 8`, *Day 6 - Using AC Powered Components* on using AC powered devices will help you to understand how to use relay devices for controlling AC powered devices; you may then apply that knowledge to control solenoid valves (as shown in the preceding figure). The operation of a solenoid valve is very simple--switching it on will open the valve and release water. While switching it off will close the valve and stop the water supply. From `Chapter 9`, *Day 7 - World of Transmitters, Receivers, and Transceivers* onwards, you will learn various wireless control and communication device techniques to use with the Arduino platform and will be able to apply them to build this smart irrigation prototype.

Try the following

At the end of this chapter, we arrive at a point, where we must apply what we have learned so far. So, let us dive into the following exercises.

- In the photodiode example, add a red LED to the circuit (refer to the `Chapter 3`, *Day 1 - Building a Simple Prototype*). Make the red LED glow as soon as the measured voltage exceeds 0.55 volts.
- In the DHT11 example, add a red LED and a green LED. Modify the `loop()` function and add logic to make the green LED glow if the temperature is below 25 degrees centigrade. Make the red LED glow if the temperature is above 25 degrees centigrade. Do not make any of the LEDs glow if the temperature is equal to 25 degrees centigrade.
- In the soil moisture sensor example, include a Piezo buzzer and make a continuous sound as soon as the measured value of soil moisture on pin A3 falls below 400.

Things to remember

Remember these important points while using the Arduino platform in your future projects:

- When using elementary sensor components, we measure the voltage generated by the sensor module using Arduino's analog pins.
- Always refer to a sensor module datasheet before interfacing it with the Arduino board

- Arduino header libraries are external C code files that can be referenced from an Arduino sketch
- Sometimes a C library must be used for an integrated sensor
- Some integrated sensor modules do not require the use of an Arduino library
- 5-volt compatible devices can be directly powered from Arduino's 5V power pin

Summary

In this chapter, we concentrated on understanding the basics of using elementary sensors as well as integrated sensor modules with the Arduino Uno board. We looked at how common electronic components such as photodiodes and photo resistors can be used for detection of light with detailed hands-on examples.

In the later section of the chapter, we learned how to use integrated sensor modules by understanding their datasheets. We learned to use a temperature sensor module and a soil moisture sensor module.

Using the basic fundamentals that you learned in this chapter, you should attempt more elementary sensors as well as integrated sensor modules on your own.

In the next chapter, we are going to learn how to build compound devices. Learning to build compound device prototypes is the first step to start building more real-world devices with the Arduino platform.

5

Day 3 - Building a Compound Device

"Compound interest is the eighth wonder of the world. He, who understands it, earns it; he who doesn't, pays it."
- Albert Einstein

We have come a long way since we started. We saw how to build basic prototypes with the Arduino in the previous chapters. In this chapter, we will take a step ahead by learning to build a compound device prototype using the Arduino platform.

We will cover the following topics in this chapter:

- Introduction to a compound device
- Using a smoke detecting sensor device
- Reading digital input from a sensor device
- Reading analog input from a sensor device
- Working with a MQ2 gas sensor
- Working with a Piezo Buzzer
- Sending digital signals to peripherals
- Receiving digital signals to peripherals
- Sending analog signals to peripherals
- Receiving analog signals to peripherals
- Read and writing using SD cards for local storage

Compound devices

A **compound device** is a collection of more than one component. The components in a compound device are inter-related to one another and centrally controlled by a micro-controller unit. Depending upon the state of one component, the other components react.

For example, a **fire alarm** is a perfect example of a compound embedded device--as soon as fire is detected, the water sprinkler system and evacuation alarm system gets activated. In the case of the fire alarm system, a fire sensor attached with a micro-controller detects fire and then the micro-controller sends signals to the peripheral components for orchestrating various activities such as activating water sprinklers and fire alarms. Advanced fire alarm systems can automatically dial emergency services via a GSM module.

Building a certified fire alarm system is an advanced topic and would be beyond the scope of this book. However, today we will build a **smoke detector device** as an example of a basic compound device. One practical application of a smoke detector can be found in non-smoking lavatories in airplanes - as soon as someone lights a smoke, the smoke is detected and the alarm goes off.

Building a smoke detector

The smoke detector device prototype will be built using a **MQ2 series gas detector module** and a **Piezo Buzzer**. The MQ2 series sensor is a general-purpose gas sensor: it can be used for sensing different types of gases (including LPG); today, we will use this sensor for detecting smoke.

As we have seen in the earlier chapters, a Piezo Buzzer is a small acoustic device that is capable of emitting sounds at various frequencies. In this example, we will write the corresponding Arduino sketch for detecting smoke and then buzzing the Piezo Buzzer five times at one second intervals. After completing `Chapter 11`, *Day 9 - Long Range Wireless Communications* which deals with GSM communications, you may use the GSM module with this chapter to send an SMS or dial a number automatically as soon as smoke gets detected. Wouldn't that be awesome!?

Apart from building a compound device prototype, both the MQ2 gas sensor and the Piezo Buzzer have been chosen in this example to demonstrate the usage of **5 volt tolerant** peripheral devices with the Arduino Uno board.

MQ2 series smoke detectors and most Piezo Buzzers commercially available in the market operate in a voltage range that can tolerate 5 volts. Therefore, both these devices can be connected directly with the Arduino Uno pins for prototyping purposes. If the external device operates at a lower voltage as compared to Arduino Uno then directly connecting the input lines of the external device may result in some damage to the lower voltage device.

Caution:
The Arduino Uno signal pins can provide a maximum current of 40 mA and the 5V pin can provide a varying range of current depending upon the power source. As a rule of thumb to start with, do not try to extract more than 200 mA from the 5V pin. Beyond these limits damage will usually start to happen to the main board. The smoke detector setup, as a whole (~160 mA for the MQ2 heater element + ~30 mA for the Buzzer + some inefficiencies in the circuit), will consume almost around 200 mA. Therefore, for long-term usage, the MQ2 smoke detector module should be powered from a separate power source to be safe, as it requires relatively large amounts of current. The example in this chapter is for short-term prototyping purpose only. If you intend to use the MQ2 smoke detector for a long period of time then do not power it from Arduino's 5V pin. You will learn how to use a separate power source in the next chapter.

On the other hand, the Piezo Buzzer has a rated maximum current of 30 mA. Hence a 100K resistor must be used as a protection while connecting the Piezo Buzzer to an Arduino Uno pin. The reason for adding a resistor was covered in `Chapter 3`, *Day 1 - Building a Simple Prototype.*

A typical MQ2 gas sensor, as shown in the following figure:

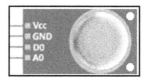

Figure 1: MQ2 gas sensor

The MQ2 gas sensor has four pins (some have three), but in this book, we will use the one that comes with four pins. The markings on the gas sensor are very obvious in this case. Pause for a moment and try to think which pins from the gas sensor should get connected to which pin on the Arduino board. Let us see how much we have understood so far.

You got it right! The Vcc pin needs to be connected to Arduino's 5V pin, whereas the GND pin should be connected to Arduino's GND pin. You may either use the digital pin D0 or the analog pin A0. The D0 pin may be connected to any digital pin on the Arduino board, while the A0 pin can be connected to any of Arduino's analog pins.

Let us try to attempt one more time and see whether we can recall how to read values for sensors. We can easily use the `digitalRead()` method to read the value on a digital pin, while we can use the `analogRead()` method to read the value on an analog pin.

Fundamental:

The input pins of a 5-volt device can be directly connected to Arduino's GPIO pins. This is because when the Arduino Uno pin transmits a signal at 5 volts, the signal can be tolerated and properly received and processed by the device attached to the Arduino.

If the external device operates at a lower voltage as compared to Arduino Uno then directly connecting the input lines of the external device may result in permanent damage to the external device.

Smoke detector - Digital I/O method

Overall, the general steps to interface the Arduino with a sensor device, as outlined in `Chapter 4`, *Day 2 - Interfacing with Sensors*, will be followed once again in this chapter. However, you will notice that for the smoke detector, we do not require an Arduino library. The reason is because the smoke detector's output can be directly read via an analog pin or via a digital pin.

For building the smoke detector project the following parts will be required:

- Arduino Uno R3
- USB connector
- 1 bread board
- 1 MQ2 gas sensor
- 1 Piezo Buzzer
- 1 pc. 100 ohms resistor
- Some male-to-male jumper wires
- Some female-to-male jumper wires

First, let us inspect the method of using the digital pin; after that we will see the method of using the analog pin. Just so you know, it may be a little difficult to trigger smoke detection using the digital pin method, compared to the analog pin method in this chapter. So, make sure the smoke density is high by using at least four to five incense sticks at a time.

Sometimes the buzzer might start beeping for the first 30 seconds, even when there is no smoke. However, after waiting for a minute, as the sensor stabilizes, it will stop beeping.

Make sure there are no loose connections in the breadboard setup. Otherwise the prototype will not work. If the buzzer still does not work, then the source of smoke should be held very close to the sensor's wire mesh.

The smoke sensor will become a little hot as it stays powered on. It is normal for the MQ2 smoke sensor to become slightly warm as it stays on. Remember not to use this setup for too long. The MQ2 gas detector needs large amounts of current and it is best to use an external power supply for long duration use.

The circuit connections for building a smoke detector prototype using its digital output pin are explained and shown in the following figure 2:

Arduino Uno pin	MQ2 gas sensor pin
Digital I/O Pin 2	D0
5V	VCC
GND	GND

Table 1: Gas sensor to Arduino (Digital I/O)

The Piezo Buzzer interfacing details are provided in the following for reference. This is the same as what we have already done in the previous chapters:

Arduino Uno pin	Piezo Buzzer pin
Digital I/O pin 8 (via 100 Ohms)	Positive terminal (longer leg)
GND	Negative terminal (shorter leg)

Table 2: Piezo Buzzer to Arduino connections

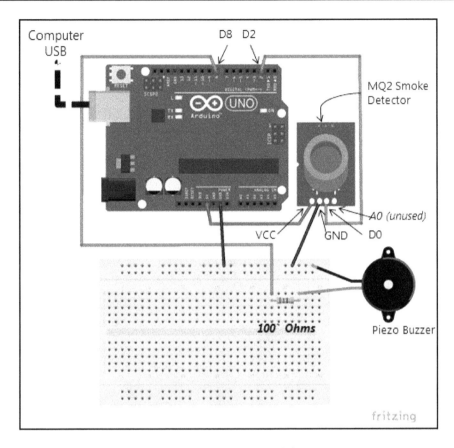

Figure 2: Smoke detector using digital I/O pins

Notice how the sensor pin D0 has been connected to Arduino Uno's digital pin number D2 for this setup for using the digital I/O method depicted in the preceding figure.

Smoke detector sketch - Digital I/O method

Use the following sketch for detecting smoke, based on the output captured from the MQ2 gas sensor's digital output pin. As always, avoid connecting the devices to the Arduino pins while loading the sketch.

The following code can be freely downloaded from the location mentioned in the *Boot Camp* section of this book. This is how the code for the smoke detector (digital I/O technique) looks:

```
//***********************************************************/
// Step-1: CONFIGURE VARIABLES
//***********************************************************/
int smokePin = 2;              // Digital Pin number 2 for
                               // connecting the smoke sensor.
int buzzerPin = 8;             // Digital Pin number 8 for
                               // connecting the buzzer.


//***********************************************************/
// Step-2: INITIALIZE I/O PARAMETERS
//***********************************************************/
void setup()
{
   pinMode(smokePin, INPUT);   // Configure Digital Pin 2
                               // in input mode,
                               // for reading signals
                               // from the smoke sensor.
   pinMode(buzzerPin, OUTPUT);// Configure Digital Pin 8
                               // for output mode,
                               // for sending signals
                               // to the buzzer.
   delay(5000);                // Wait for 5 seconds
                               // before sending any signals
                               // to the sensor.
                               // Although it is not required
                               // by the smoke sensor's
                               // datasheet;
                               // However, it is a good
                               // practice to let the
                               // sensor device stabilize
                               // for a few seconds before
                               // sending any signals to it.
}


//***********************************************************/
// Step-3: MAIN PROGRAM
//***********************************************************/
void loop()
{
  // Read the input at Arduino Uno's Digital Pin 8.
  // As soon as smoke is detected:-
  // (1) The onboard LED on the MQ2 gas sensor will glow, and
  // (2) a LOW signal will be received on Digital Pin 8
  if (digitalRead(smokePin) == LOW)
```

```
    {
      // Ring the buzzer 5 times at 30 second intervals.
      tone(buzzerPin, 500, 250);
      delay(500);
      tone(buzzerPin, 500, 250);
      delay(500);
      tone(buzzerPin, 500, 250);
      delay(500);
      tone(buzzerPin, 500, 250);
      delay(500);
      tone(buzzerPin, 500, 250);
      delay(500);
    }
    // Wait for 2 seconds and check again
    delay(2000);
  }
```

The preceding sketch is a very simple example of reading the digital output received on Arduino Uno's digital pin number 8. It is important to note the technique used to read the digital input signal from the sensor. Let us try to understand the sketch in detail. The sketch starts with the typical section on declaring variables to be used in the sketch. In this case, we have defined two variables: the smokePin and the buzzerPin.

The setup() function is used to configure the pins used in this setup in the appropriate mode. Digital pin number 2, which is connected to the D0 pin of the MQ2 sensor module, is configured in input mode, so that it can be used to read the input signal from the MQ2 smoke sensor.

Digital pin number 8 is configured in output mode so that it can be used to send digital output signal via digital pin number 8. Then there is a small delay of 5 seconds. This is not mandatory, but it has been done to allow the smoke sensor to stabilize, as shown in the following code:

```
    delay(5000);                    // Wait for 5 seconds
```

The final step is to use Arduino's in-built function digitalRead(<pin-number>) for reading the input signal level (HIGH/LOW) available on pin 8 (in this example). The digitalRead(<pin-number>) function takes the pin number as an input and returns the signal level available on the digital input pin.

In this example, it checks the signal level on pin number 8 and returns the signal level. A LOW signal level is received as soon as smoke is detected. In conjunction, it would be good to note that there is an onboard LED marked as D0-LED on the reverse side of the MQ2 gas sensor circuit board: this LED starts glowing every time smoke is detected:

- **Receiving digital signals**: Use Arduino Uno's digital I/O pins for connecting to a digital I/O pin of a peripheral device. Configure Arduino's digital I/O pin in input mode, for reading input signals from a peripheral device. Use the `digitalRead(<pin-number>)` function for programmatically reading the input signal, at the specified pin number. The use of this method is depicted in the preceding smoke detector example.
- **Sending digital signals**: Configure Arduino's digital I/O pin in OUTPUT mode, for sending output signals to a peripheral device. Use the `digitalWrite(<pin-number>, <HIGH/LOW>)` function for programmatically sending output signals to a peripheral device, via the specified pin number.

The second part of the preceding sketch focuses upon sounding the Piezo Buzzer five times, at intervals of 30 seconds. Arduino's in-built function is used to emit sounds using the Piezo Buzzer:

```
tone(buzzerPin, 500, 250);
```

In this example, the function call causes the Buzzer to emit a sound at a frequency of 500 Hertz for a period of 250 milliseconds. While the `delay(500)` function simply delays the program execution for 500 milliseconds (that is, 30 seconds).

Sometimes due to unspecified or unexpected conditions, the electronic peripheral sensor devices might require a few extra seconds to stabilize. Unless specified otherwise in the sensor device's datasheet, it is recommended to halt the program execution for a few seconds in the `setup()` function, before sending any signals to the peripheral sensor devices. For example, in the preceding sketch, the program execution has been halted for 5 seconds, before any interaction with the MQ2 gas sensor starts.

Smoke detector (analog I/O method)

Now let us examine the method of reading the analog output pin of the MQ2 gas sensor. The circuit connections for building a smoke detector prototype using its analog output pin is shown in *figure 3*. As shown in the figure, the MQ2 gas sensor to Arduino Uno connection, via its analog pin, is tabulated as follows:

Arduino Uno pin	MQ2 gas sensor pin
Analog I/O pin A5	A0
5V	VCC
GND	GND

Table 3: MQ2 gas sensor to Arduino - Analog connections

The remaining portions of the circuit, such as the Piezo Buzzer connection, will remain unaltered from the previous example with the digital pin. The Piezo Buzzer interfacing details are provided in the following table for reference.

Arduino Uno Pin	Piezo Buzzer Pin
Digital I/O pin 8 (via 100 Ohms)	Positive terminal (longer leg)
GND	Negative terminal (shorter leg)

Table 4 - Piezo Buzzer interfacing

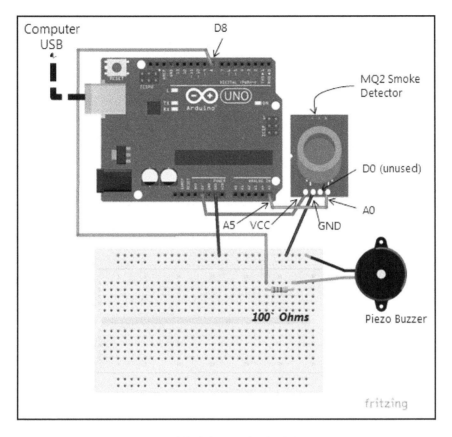

Figure 3: Smoke detector (analog I/O method)

Notice how the sensor pin A0 has been connected to Arduino Uno's analog pin number A5 for this setup using the analog I/O method.

Smoke detector sketch (analog I/O method)

The following sketch should be used for detecting smoke, based on the output captured from the MQ2 gas sensor's analog output pin. As always, avoid connecting the devices to the Arduino pins while loading the sketch.

The following code can be downloaded from the location mentioned in Chapter 1, *Boot Camp*. Here is the code for the smoke detector (analog I/O technique):

```
//********************************************************/
// Step-1: CONFIGURE VARIABLES
//********************************************************/
int smokePin = A5;            // Analog Pin number A5
                              // for connecting the smoke
                              // sensor.
int buzzerPin = 8;            // Digital Pin number 8
// for connecting the buzzer.

//********************************************************/
// Step-2: INITIALIZE I/O PARAMETERS
//********************************************************/
void setup()
{
  pinMode(smokePin, INPUT);   // Configure Analog Pin A5
                              // in input mode,
                              // for reading signals
                              // from the smoke sensor's
                              // Analog output pin.

  pinMode(buzzerPin, OUTPUT); // Configure Digital Pin 8
                              // for output mode,
                              // for sending signals
                              // to the buzzer.

  delay(5000);                // Wait for 5 seconds
                              // before sending any signals
                              // to the sensor.
                              // Although it is not required
                              // by the smoke sensor's
                              // datasheet;
                              // it is a good practice
                              // to let a sensor device
                              // stabilize
                              // for a few seconds before
                              // sending any signals to it.
}

//********************************************************/
// Step-3: MAIN PROGRAM
//********************************************************/
void loop()
{
  // Read the input at Arduino Uno's Analog Pin A5.
  // As soon as smoke is detected
```

```
  // the reading should be above 325
  if (analogRead(smokePin) > 325)
  {
    // Ring the buzzer 5 times at 30 second intervals.
    tone(buzzerPin, 500, 250);
    delay(500);
    tone(buzzerPin, 500, 250);
    delay(500);
    tone(buzzerPin, 500, 250);
    delay(500);
    tone(buzzerPin, 500, 250);
    delay(500);
    tone(buzzerPin, 500, 250);
    delay(500);
  }

  // Wait for 2 seconds and check again
  delay(2000);
}
```

The preceding sketch is a very simple example of reading the analog output received on Arduino Uno's analog pin number A0. It is important to note the technique used to read the analog input signal from the sensor.

The first step is to connect the sensor's analog output pin to Arduino Uno's analog I/O pin number A0 (in this example). The next step is to configure (via the C sketch) Arduino's pin A0 in INPUT mode. The final step is to use Arduino's in-built function - analogRead(<pin-number>) - for reading the input signal available on pin A0 (in this example).

The integer number 325 used for detecting the smoke threshold might need some calibration based on the type of smoke being detected. Hence it is advisable to adjust the number in order to detect a particular type of smoke.

The `analogRead(<pin-number>)` function takes the pin number as an input and returns an analog value corresponding to the signal available on the analog input pin. In this example, it checks the analog signal available on pin number A0 and returns an integer value corresponding to the signal on the A0 pin. This integer number is usually greater than 325, as soon as smoke is detected:

- **Receiving analog signals**: Use Arduino Uno's analog I/O pins for connecting to an analog I/O pin of a peripheral device. Configure Arduino's analog I/O pin in INPUT mode, for reading input signals from a peripheral device. Use the `analogRead(<pin-number>)` function for programmatically reading the analog input signal at the specified pin number. The use of this method is depicted in the preceding smoke detector example.
- **Sending analog signals**: Configure Arduino's analog I/O pin in OUTPUT mode, for sending analog output signals to a peripheral device. Use the `analogWrite(<pin-number>, <value>)` function for programmatically sending analog signals to a peripheral device, via the specified pin number.

Now that we have seen how a basic compound device is built, the next step would be to introduce another useful dimension while prototyping compound devices. In the real world, most devices have the need to store data. For example, an internet-enabled device has the capability to store your Wi-Fi credentials. Another practical example would be a conference room light controller that has the ability to be configured with user preferences such as switching off the conference room light if there is no movement in the room for an hour. Similarly, there are a myriad of uses for storing data locally in a device.

In the next section, we will introduce the usage of local storage, using the SD card module. We will learn how to use the **SD card module** by adding the module to the smoke detector device that we built earlier in this chapter. Usually, large amounts of data get logged by sensors and therefore SD cards provide an inexpensive temporary local storage, until the data is flushed to IoT Cloud channels. `Chapter 12`, *Day 10 – The Internet of Things Project* will explain what IoT cloud channels are.

It is worthwhile to note that the ATmega328 microcontroller on the Arduino board comes in-built with a local data storage of 1024 bytes, for storing very small amounts of data. This memory area is known as EEPROM (electrically erasable programmable read-only memory). The data stored in the EEPROM is persisted even after switching off the power supply to the board. You can refer to the following Arduino Foundation website URL for a tutorial of writing data to the EEPROM area:

`https://www.arduino.cc/en/Tutorial/EEPROMWrite`.

Upon rebooting the Arduino board we can easily retrieve the stored information. You can refer to the following Arduino foundation website URL for a tutorial of retrieving data from the EEPROM area:

`https://www.arduino.cc/en/Tutorial/EEPROMRead`.

Local storage with SD card modules

SD card storage is a relatively wide subject. Therefore, we will focus on the basics of interfacing, reading, and writing data to and from a **micro SD card** with the Arduino Uno platform. At the end of this topic, you will clearly understand how SD cards can be used in any electronic device for local storage capabilities.

In the example that we will build, our focus will be to understand the following basic activities:

- Reading pre-configured data from an SD card and then using it in the C sketch
- Writing new data to local files that can be ported or transferred to other mediums as required

Reading pre-configured data from an SD card is a common technique to have an external place to store the values of variables that are used in the C sketch. This is a normal practice to read system configurations from a local storage device. The variables in the C sketch in turn read these stored values from the files on the SD card and can apply them during the execution of the sketch. The main advantage of being able to do so is that the values of the variables can be modified just by changing the values in the external files in the SD cards. This allows the device user to set device preferences as per their needs, instead of hard-coded configurations in C sketch.

Being able to record new data locally in files is a very common practice for devices that frequently generate data that cannot be, or should not be, transmitted to central databases over an internet connection. For example, imagine a situation where a temperature data logger device is located in a freezer inside a research laboratory. The temperature of the freezer has to be monitored regularly and transmitted to a central database. There might be unforeseen situations when the internet connection might get disrupted and the temperature data logger device will no longer be able to transmit the data to the central server. During this period when the data logger remains offline (not connected to the internet), the data to be transmitted is normally safely stored on a local storage medium (such as an SD card). As soon as internet connectivity is established, the data is read from the local storage medium and posted to the central server.

For learning to use the SD card module, we will build upon our setup of the previous smoke detector using the analog I/O method. Note that in this new circuit, the smoke detector has been powered from the common power rail on the breadboard, instead of directly plugging it into the 5V pin.

Our objective in this lesson would be to do the following two things:

- Set the value of the `samplingInterval` variable from a pre-configured value in a file on the SD card
- Whenever smoke is detected, we will store the values in a file on the SD card

Apart from the parts that we have already used in this chapter, the following additional parts will be required following through the remainder of this chapter:

- 1 SD card module (with in-built 5V to 3.3V logic shifter; most modules have this. If in doubt, ask your seller and also check whether an AMS 1117 3.3V voltage converter chip is present on the module's PCB)
- 1 SD card
- Six pieces of female-to-male jumper wire

This book uses a micro SD card with a micro SD card module to be precise. However, you may use SD cards of other sizes as well. However, in that case, a matching SD card module must be used. Refer to the following breadboard setup for wiring the SD card module with your device prototype.

SD card interfacing and communication is based on the **Serial Peripheral Interface** (SPI) communication technique. In the earlier days, SPI communication was prevalent for connecting peripheral devices such as printers and scanners to a computer, however, this has mostly been replaced by USB communication. Nowadays, SPI communication is mostly used to control peripheral devices such as memory cards, display units and sensors. As part of SPI communications a pair of devices are connected together - one being the master device while the other is the slave device. Using this technique data exchange can take place in full-duplex (that is, in both directions) between the master and the slave device. In our case the Arduino will be the master device and the SD Card module will be the slave device. Although SD card storage is an advanced topic, for our knowledge it would be good to know what the various pins on the SD card module are. The pins are described in the following:

- **CS/SS (Chip Select/Slave Select)**: This pin is used to enable the SD card for use. The CS/SS pin can be connected to any available digital pin (in this book we will use digital pin 4).

- **SCK/CLK (System Clock)**: This line synchronizes with the Arduino clock pulses for data transfer. In our circuit, this pin will be connected to Arduino's digital pin 13.
- **MOSI (Master Out Slave In)**: Line for sending data from the Arduino board to the SD card. The Arduino Uno board is the master device and the SD card is the slave device. In our circuit, this pin will be connected to Arduino's digital pin 11.
- **MISO (Master In Slave Out)**: Line for reading data from the SD card into the Arduino board. In our circuit, this pin will be connected to Arduino's digital pin 12.
- **VCC**: For power supply to the SD card module.
- **GND**: For ground connection.

 The digital pins to be used for interfacing the Arduino Uno and the SD card module are pre-configured in the SD card library that will be used. So, you will not be able to change the pins. The only exception is the SD card module's CS/SS pin that can be connected to any available digital pin on the Arduino board.

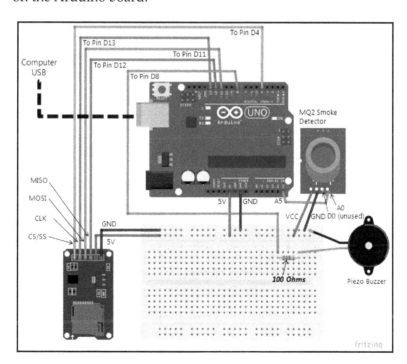

Figure 4: Wiring of SD card with Arduino-based compound device

The details of the connections between the SD card module and the Arduino Uno are provided in the following table for easy reference:

SD card module pin	Arduino Uno pin	Connection comment
CS/SS	Pin D4	Connect directly
SCK/CLK	Pin D13	Connect directly
MOSI	Pin D11	Connect directly
MISO	Pin D12	Connect directly
VCC	5V	Connect via breadboard rail
GND	GND	Connect via breadboard rail

Table 5: SD card to Arduino Uno connections

The details of connecting the smoke sensor and the Arduino Uno are provided in the following table for easy reference:

Smoke sensor pin	Arduino Uno pin	Connection comment
VCC	5V	Connect via breadboard rail
GND	GND	Connect via breadboard rail
A0	Pin A5	Connect directly

Table 6: Smoke Sensor to Arduino Uno connections

The details of connecting the Piezo Buzzer and the Arduino Uno are provided in the following table for easy reference:

Smoke sensor pin	Arduino Uno pin	Connection comment
Positive terminal (longer leg)	Pin D8	Connect via 100 ohms resistor
Negative terminal (shorter leg)	GND	Connect via breadboard rail

Table 7: Piezo Buzzer to Arduino Uno connections

After building the above circuit, before proceeding with loading the main sketch, make sure that you have created a text file titled `IntStng.txt` on the SD card as explained in the following. This file will in turn be used in the main sketch. Follow the following instructions carefully; otherwise the sketch will not work:

- The file name should be `IntStng.txt`
- There should be only one number (between 1-9) written in the first line of the file
- Ensure that the file is in the root folder on the SD card

You may use Windows Explorer to access the SD card and create this file or you may download the file from the online location for this chapter mentioned in the boot camp chapter of this book, and then upload the file onto the SD card. The file should look like what is shown in the following figure. This screenshot was taken via Notepad on a Windows 7 computer:

Figure 5: Settings file on SD card

Using our sketch, we will read the preceding pre-configured value and then use it in the sketch. Load the following sketch into your Arduino board. You may download the sketch from the location mentioned in the boot camp chapter of this book. After loading the sketch, launch the serial monitor window and see how the program executes. You may add your own `Serial.println()` statements at appropriate places in the sketch to view additional output.

The following code shows us how to use an SD card with smoke detector device:

```
//********************************************************/
// Step-1: CONFIGURE VARIABLES
//********************************************************/
#include <SPI.h>              // For SD Card
#include <SD.h>               // For SD Card
// declare handles to the data logging file
File logFile;           // log file to which smoke detected events
                        // will be logged
File settingsFile;      // contains interval setting after which
                        // smoke detection function should be run
int smokePin = A5;
int buzzerPin = 8;
```

```
// 2 seconds by default, but will be changed through the value
// configured in the IntStng.txt file in the SD Card
int samplingInterval = 2000;
bool sdCardOK = false;

//**********************************************************/
// Step-2: INITIALIZE I/O PARAMETERS
//**********************************************************/
void setup()
{
  // Initialize serial communication
  // wait for serial monitor window to connect via USB port
  Serial.begin(9600);
  while (!Serial) { ; }

  // Initialize the SD card
  // The SD object used below is available globally
  // The SD object is defined in the header files
  // Digital Pin 4 is used to enable the SD card
  if(!SD.begin(4))
  {
    Serial.println("There seems to be a problem with the SD card");
    Serial.println("Default sampling interval will be 2 seconds");
  }
  else
  {
    sdCardOK = true;
    // Read the sampling interval from SD Card
    samplingInterval = readSamplingIntervalSetting("IntStng.txt");
    Serial.print("Sampling Interval Found from SD Card = ");
    Serial.println(samplingInterval);
  }

  pinMode(smokePin, INPUT);
  pinMode(buzzerPin, OUTPUT);
}

//**********************************************************/
// Step-3: MAIN PROGRAM
//**********************************************************/
void loop()
{
  // Read the input at Arduino Uno's Analog Pin A5.
  // As soon as smoke is detected, the reading should be above 325
  int value = analogRead(smokePin);
  if (value > 325)
  {
    Serial.println("Smoke detected...");
```

```
    // Log the event to the SD Card
    // data will be logged only if SD Card is available properly
    if(sdCardOK)
    {
      logDataToFile(value);
    }
    else
    {
      Serial.println("SD Card not proper, data not logged.");
    }

    // Ring the buzzer 5 times at half second intervals.
    tone(buzzerPin, 2000, 1000);
    delay(1000);
    tone(buzzerPin, 2000, 1000);
    delay(1000);
    tone(buzzerPin, 2000, 1000);
    delay(1000);
    tone(buzzerPin, 2000, 1000);
    delay(1000);
    tone(buzzerPin, 2000, 1000);
    delay(1000);
  }

  // Wait for the number of seconds as per settings file
  // in the SD card file IntStng.txt
  delay(samplingInterval);
}

void logDataToFile(int data)
{
  // open the data logging file
  logFile = SD.open("datalog.txt", FILE_WRITE);

  // if the log file opens successfully
  if (logFile)
  {
    // then write the measured parameters in the log
    logFile.println(data);
    Serial.println("Data logged");
    // the file should be closed after data logging
    logFile.close();
  }
  else
  {
    Serial.println("File datalog.txt could not be opened");
  }
}
```

```
int readSamplingIntervalSetting(String fileName)
{
  // This function will read and display pre-configured
  // settings from a settings file on the Serial Monitor window
  // Assumption: IntStng.txt file has only 1 line
  // and in that first line only an integer value is expected
  settingsFile = SD.open(fileName, FILE_READ);
  int iSetting = 0;

  // if the file was opened successfully
  if (settingsFile)
  {
    // read from the file one line at a time
    // this will go on until it reaches the end of the
    // file
    int counter = 0;

    // this loop is designed to read only the
    // first character on the first line in the file
    while (settingsFile.available())
    {
      counter = counter + 1;
      iSetting = settingsFile.read();
      if(counter > 0) //first character has been read
      {
        break;    //break out of this while loop
      }
    }
    iSetting = 3 * 1000; // convert to milli seconds

    // close the file after reading is complete
    settingsFile.close();
  }
  else
  {
    // the file could not be opened
    Serial.println("Could not open IntStng.txt");
  }
  return iSetting;
}
```

Let's understand the important parts of the sketch. At the beginning of the sketch, two header files have been included. These header files are required for interfacing with the SD card properly. Basically, these header files work as the driver software for interfacing with the SD card:

```
#include <SPI.h>          // For SD Card
#include <SD.h>           // For SD Card
```

The next important lines of code are the file handles (also known as file pointers). These file handles are declared at the top of the sketch and are referred to throughout the sketch, wherever there are file operations:

```
File logFile;
File settingsFile;
```

Thereafter, the `setup()` function initializes all the necessary pins and parameters for the sketch. First of all, the SD card is initialized by using the following function call. The hardcoded value 4 indicates Arduino's digital pin number 4. The SD card is enabled using this pin number 4:

```
SD.begin(4)
```

The `SD` object is available globally via the header files. The `SD` object provides a reference to the SD card attached to the Arduino Uno board. Various functions can be invoked using the `SD` object.

After the SD card gets initialized, the value of the `samplingInterval` variable is fetched from a file on the SD card. The `samplingInterval` variable is used to specify a default sampling interval that will be used in the `delay(samplingInterval)` function at the end of the `loop()` function, so that the sketch checks the smoke sensor readings at regular intervals. The default value is replaced by the value present in the `IntStng.txt` file in the SD card. The code in the `readSamplingIntervalSetting()` function is used to read the value from the `IntStng.txt` file on the SD card.

The function `readSamplingIntervalSetting(String fileName)` starts with a statement to open the file as shown in the following code:

```
settingsFile = SD.open(fileName, FILE_READ);
```

In the preceding line of code, the `SD.open()` function has to be called by passing the name of the file to be opened and the `FILE_READ` parameter to specify that a file needs to be opened in read mode.

After opening the file in read mode, the first character in the first line of the file is extracted using the following `while` loop logic:

```
while (settingsFile.available())
{
  counter = counter + 1;
  iSetting = settingsFile.read();
  if(counter > 0) //first character has been read
  {
    break;          //break out of this while loop
  }
}
iSetting = 3 * 1000; // convert to milli-seconds
```

So, as per the preceding extraction logic, let us say you want to increase the sampling interval to 5 seconds; then simply ensure that the first character in the first line of the `IntStng.txt` file is 5. The logic will multiply the extracted number by 1000 to convert it to milli-seconds, for use with the `delay()` function. Similarly, you may specify any other integer in the `IntStng.txt` file to specify the sampling interval in seconds. The following function call helps read the characters from the file:

```
settingsFile.read();
```

After the setting has been extracted, the file must be closed by calling the `close()` function on the settings file handle:

```
settingsFile.close();
```

Only one file can be opened at a time. It is mandatory to close the file once reading and or writing has been done on it.

The code in the `loop()` function is straightforward. It builds upon the previous smoke detector example. After the condition to detect smoke is satisfied, the following function is called:

```
logDataToFile(buf);
```

First the preceding function opens the data logging file on the SD card by calling the following file opening function on the global `SD` object:

```
logFile = SD.open("datalog.txt", FILE_WRITE);
```

After the file has been opened in write mode by specifying the `FILE_WRITE` parameter, the data is logged into the file by using the following line. For the first time, if the file does not exist, then a new file will be created and the data will get saved as the first line of the new file. From the second time onwards, the data will be logged after the existing last line in the file:

```
logFile.println(data);
```

In the end, once again the file must be closed. After running the prototype for some time and after logging some data by simulating smoke detected events, you can view the contents of the `datalog.txt` file from Windows Explorer. You would see something like the following figure:

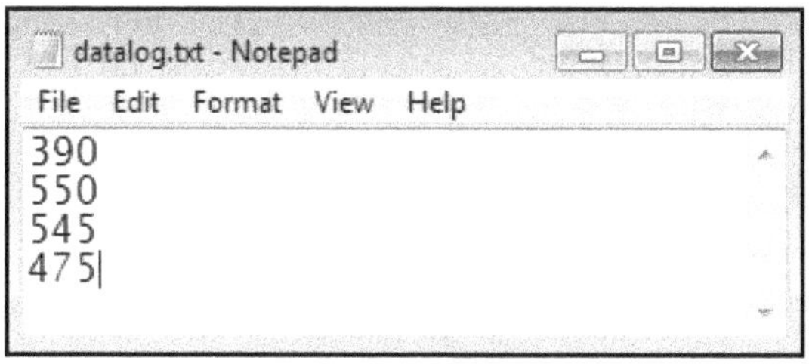

Figure 6: The datalog.txt file

With this, we come to the end of Day 4 of our journey into the world of device prototyping with the Arduino platform. In the next chapter, we will build one more compound device prototype, however, this time we will take a leap further and get introduced to the concept of building a battery powered device, thus creating our first standalone device.

Try the following

Now let us perform some exciting exercises for a better understanding of what we have learnt so far.

- Refer to the `Chapter 3`, *Day 1 - Building a Simple Prototype*; add an external LED to the circuit via a digital pin.
- Modify the sketch and make the external LED also blink along with sounding the buzzer.
- Modify the sketch, to perform the buzzing and blinking using a `for` loop.

- In the SD card example, try creating two log files instead of only one `datalog.txt`. One file should be called `critical.txt` and should log events where the A5 pin reading is above 500. While all other readings should continue to get logged in the existing `datalog.txt` file.

- Write a sketch to list the contents of an existing file by adding the following function in the sketch. You can invoke this function by using a function call such as `readFullFile("datalog.txt")` from the `loop()` function of the sketch:

```
void readFullFile(String fileName)
{
  // This function will read and display the
  // file contents in the Serial Monitor window
  // open the file for reading the already logged values
  logFile = SD.open(fileName, FILE_READ);

  // if the file was opened successfully
  if (logFile)
  {
    // read from the file one line at a time
    // this will go on until it reaches the end of the
    // file
    while (logFile.available())
    {
      Serial.write(logFile.read());
    }

    // close the file after reading is complete
    logFile.close();
  }
  else
  {
    // the file could not be opened
    Serial.print("Could not open file");
    Serial.println(fileName);
  }
}
```

- Change the connection of the CS/SS pin with Arduino's digital pin 4 to digital pin 5 and also change the following, corresponding line of code in the `setup()` function. This statement will enable the SD card to use digital pin 5. Similarly, other digital pins may also be used:

```
SD.begin(5);
```

Things to remember

Remember the following things. As you move forward, you will get your own ideas based on these points so that you can start creating more of your own devices.

- A 5-volt peripheral device can be directly connected to Arduino's 5V power pin.
- All the peripheral components must have a common ground.
- The `analogRead` function is used to read input from an analog pin.
- The `analogWrite` function is used to send output signals to an analog pin.
- The `digitalRead` function is used to read input from a digital pin.
- The `digitalWrite` function is used to send output signals to a digital pin.
- You will notice that the intensity of the Piezo Buzzer decreases as more peripheral devices are added. This is because the number of devices consuming power from the same 5V pin of the Arduino board have increased.
- The digital pins used to interface with an SD card are pre-configured in the SD libraries used with Arduino. Hence you will not be able to change the pin numbers for MOSI, MISO, and CLK/SCK.
- However, the connection to the CS/SS pin maybe changed. It can be connected to any available digital pin on the Arduino board.
- Using the preceding information, more than one SD card may be interfaced with the Arduino board. In which case, the SD cards will have separate CS/SS lines but share all other lines. Using this technique, you may implement a backup SD card in case the primary SD card becomes unavailable or corrupt.

Summary

In this chapter, we learnt how to build a compound device using the Arduino board as the controlling device. We saw how a compound device can be easily built around the Arduino board by connecting multiple peripheral devices.

The MQ2 gas sensor that we used in this chapter can be used in a variety of real world applications such as, household gas leakage detection, handheld gas detectors for geographical surveyors and underground workers, and so on. Similarly, you can now start imagining numerous compound devices that can be built around the Arduino board.

In the next chapter, we will learn how to create a standalone device that is not connected to the computer and that runs on a DC battery power source. If you look around, almost all devices are of a standalone nature. They are powered from a DC battery source and can run in a self-contained manner. Thus, we will move a step closer to our journey of learning Arduino-based device prototyping.

6

Day 4 - Building a Standalone Device

"Be sure you put your feet in the right place, then stand firm."

- Abraham Lincoln

So far, in all the previous chapters, we have been building and powering the prototypes by plugging in the Arduino board into the computer using a USB cable. First, the Arduino board would get powered by this USB connection. In turn, the Arduino board was supplying power to the connected peripheral devices.

While building real-world device prototypes, independent (not from a computer's USB port) power sources must be used, so that the device prototype can work without being connected to a computer. Hence, in this chapter, we will learn how to make standalone devices that have their independent power sources.

You will learn the following topics in this chapter:

- Introduction to standalone devices
- Different considerations for power sources based on the type of project
- Building a distance measurement gadget
- Using an LCD panel
- Using battery power
- Using HC-SR05 ultrasonic sensor

Standalone devices

We are already familiar with the basic parts required for building a prototype. The two new aspects to consider when building a standalone project are an independent power source and a project container.

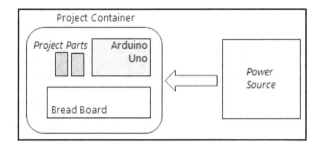

Figure 1: Typical parts of a standalone prototype

As shown in the preceding diagram, typically a standalone device prototype will contain the following parts:

- The device prototype (Arduino board + peripherals + all the required connections)
- An independent power source
- A project container/box

After the basic prototype has been built, the next consideration is to make it operable on its own, like an island. This is because in real-world situations, you will often have to make a device that is not directly connected to and powered from a computer. Therefore, we will need to understand the various options that are available for powering our device prototypes and also understand when to choose which option.

The second aspect to consider is an appropriate physical container to house the various parts of a project. A container is a physical device container, which will ensure that all parts of a project are nicely packaged in a safe and aesthetic manner.

External power supply options

The Arduino Uno board has three different options for receiving power:

- USB B port
- DC IN power socket
- VIN pin

Providing power via the USB B port in a standalone manner can be achieved via USB power hubs. These USB power hubs are also sometimes called USB power banks. The USB B port can receive 5 volts of power.

The DC IN power socket can receive power supply from external batteries as well as AC to DC wall adapters. The DC IN socket can receive 7-12 volts of power supply.

The VIN pin provides yet another option for receiving power from external power supplies. In this case, mostly, external batteries are used. Similarly, the VIN pin can also receive 7-12 volts of power.

 As a caution, avoid an input voltage on the 5V pin, as the input voltage to the 5V pin will bypass the on-board voltage regulator and enter the main board circuit directly. This will be a problem if a voltage higher than 5 volts (say 9-12 volts) is accidentally connected to the 5V pin. Although a regulated 5 volts can be input through the 5V pin, avoid it as a precaution.

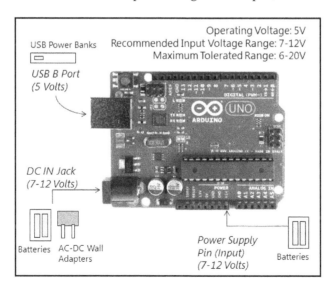

Figure 2: Power supply options to the Arduino Uno board

When using the USB B port and the DC IN jack, simply plug in the USB B connector or the DC barrel jack into the appropriate socket on the Arduino board. However, if the VIN pin is chosen to provide power, then the positive terminal of the power source must be plugged into the VIN pin, while the negative terminal should be plugged into a GND pin of the Arduino board.

Now that we are aware of the various power supply options available to us, let's understand the rationale behind choosing them depending upon our project requirements.

The various popular options for powering Arduino projects in a standalone manner are tabulated here for reference:

Power source	Arduino Uno input port	Suitability for project types	Comments/Constraints
AC to DC wall adapter	DC IN Socket	Projects requiring constant and stable supply	Proximity to electric power supply lines
Alkaline batteries	DC IN Socket or VIN Pin	Projects requiring short and intermittent supply	Power stability is comparatively lower than Lithium Ion batteries
Lithium Ion batteries	DC IN Socket or VIN Pin	Projects requiring short and intermittent supply	Power stability is comparatively better than Alkaline batteries
USB power hub	USB B Port	Projects requiring short and intermittent supply	Power stability is comparatively better than Alkaline batteries

Table 1: Power source considerations

Apart from the power supply methods mentioned earlier, there are other unconventional methods such as drawing power from Solar charged cells as well. However, these techniques are beyond the scope of this book.

Sometimes, due to the number of parts and type of parts used in a prototype, you may have to use more than one power source in a single project. This is because a single power source may not be enough for catering to the power requirements of all the combined parts. We will see a practical example in Chapter 7, *Day 5 - Using Actuators*.

Determining power source capacity

This topic is one of the trickiest topics. There is no one size which fits all solution. The power requirements for the same device may vary depending upon how it is used by the sketch and how the circuit has been designed. Power source capacity is usually measured in terms of how much current can be supplied by a battery and for how long it can supply the current. Precise power consumption is an advanced electrical engineering topic; we will focus on learning how to determine approximately how many batteries we should use in our projects. Learning this should help you to become ready for the challenge of building standalone devices.

How many batteries will my device prototype need? How long will it run given x number of batteries attached to it? Let's try to understand the answers to these questions. The power source capacity is usually measured in terms of how much current it can supply over a finite period of time. The unit to the measure capacity of a battery is mAh (milli-Ampere-hour). 1000 mAh means a battery can supply a current of 1000 milli-amperes for 1 hour. In other words, if your device requires 1000 mA of current continuously, then a battery of 1000 mAh can last only for 1 hour.

Therefore, depending upon how much current is required by the device prototype and for how long, we will need to calculate the capacity of an independent battery power source. Any prototype that we build will have multiple parts. Each part will have its power rating in terms of current consumption. Based on the total power rating of all the combined parts, you will have to determine the type and capacity of the power source to be used.

Let's take a very basic example of blinking a Red LED with the Arduino Uno R3 board. Assume that the Red LED should be blinked at 1 second periods, for 3600 times over a 24 hour period - so, effectively, the Red LED would be blinked for 3600 seconds, which equates to 1 hour. From a Red LEDs datasheet, we know that the Red LEDs consume around 20 mA (on the higher side). Therefore, the amount of current drawn by the Red LED would be calculated as:

Current consumption = Current Rating * Blink Duration * Number of blinks per day

= 20 mA * 1 second * 3600 times per day

= 20 mA * 3600 seconds

= 20 mA * 1 hour => 20 mAh per day

Therefore, the red LED alone would draw 20 mAh of current/energy from a battery every day.

Similarly, we will need to calculate the current/energy drawn by all the other parts in our device prototype. In this example, the only other part is the Arduino Uno R3 board, which has a variable current draw between 20-50 mA depending upon various onboard parts being in use or not being in use and the operation mode it is in. For the sake of understanding the battery capacity topic, let's assume that the Arduino Uno board consumes an average of 40 mA and it remains operational throughout the day (without being sent into sleep mode); then the current/energy drawn by the Arduino Uno R3 board would be calculated as:

Current consumption = Average Current Consumption * Number of hours per day

= 40 mA * 24 hours per day

= 960 mAh per day

Therefore, the Arduino Uno R3 board itself would draw 960 mAh of current/energy from a battery every day.

So, as a whole, the energy/power consumption of the device prototype, on a per day basis would be calculated as:

Total current consumption = Arduino Uno R3 + Red LED

= 960 + 20 mAh per day

= 980 mAh per day

Arduino's DC IN port recommends a voltage of 7-12 volts; therefore, we would need at least 5 AA sized 1.5V batteries connected in series to achieve a voltage of 7.5 volts.

1 AA sized 1.5V alkaline battery rated capacity = 1800-2600 mAh ~ 2000 mAh

So 5 AA sized 1.5V batteries would be able to supply = 10000 mAh

Therefore, the number of days the 5 pack battery would last = Total energy / Daily need

= 10000 / 960 days

~ 10 days

This is how you can determine the approximate power source requirements for your device prototype. You must also remember that the preceding calculation is tricky because there are a lot of variables at play, such as whether the Arduino is in sleep mode or not. The additional current will also be lost as a result of dissipation, wire lengths, and so on. It is usually a good practice to err on the side of caution and provide for an extra 10-15% capacity than the calculated value.

Now, let's start exploring how to build a standalone device using the Arduino platform in the following sections of this chapter.

Building a distance measurement device

Let's build an exciting project by combining an Ultrasonic Sensor with a 16x2 LCD character display to build an electronic distance measurement device. We will use one of the most easily available 9-volt batteries for powering this standalone device prototype.

For building the distance measurement device, the following parts will be required.

- Arduino Uno R3
- USB connector
- 1 pc. 9 volt battery
- 1 full sized bread board
- 1 HC-SR04 ultrasonic sensor
- 1 pc. 16x2 LCD character display
- 1 pc. 10K potentiometer
- 2 pcs. 220 ohms resistor
- 1 pc. 150 ohms resistor
- Some jumper wires

Before we start building the device, let's understand what the device will do and the various parts involved in the device. The purpose of the device will be to be able to measure the distance of an object from the device. The following diagram depicts the overview of the device:

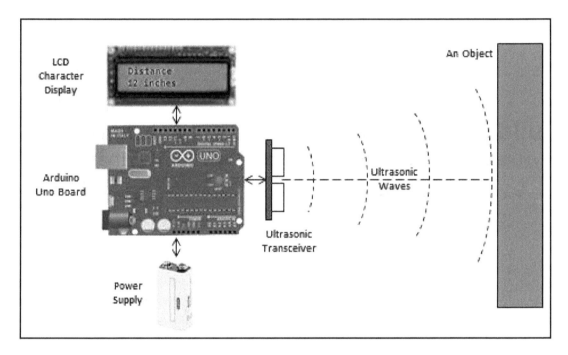

Figure 3 - A standalone distance measurement device overview

First, let's quickly understand each of the components involved in the preceding setup. Then, we will jump into hands-on prototyping and coding. The ultrasonic sensor model used in this example is known as **HC-SR04**. HC-SR04 is a standard commercially available ultrasonic transceiver.

A transceiver is a device that is capable of transmitting as well as receiving signals. The HC-SR04 transceiver transmits ultrasonic signals. Once the signals hit an object/obstacle, the signals echo back to the HC-SR04 transceiver. The HC-SR04 ultrasonic module is shown below for reference.

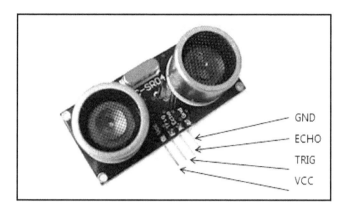

Figure 4: The HC-SR04 Ultrasonic module

The HC-SR-04 has four pins. The usage of the pins is explained below for easy understanding:

- **Vcc**: This pin is connected to a 5 volt power supply
- **Trig**: This pin receives digital signals from the attached microcontroller unit in order to send out an ultrasonic burst
- **Echo**: This pin sends the measured time duration proportional to the distance travelled by the ultrasonic burst
- **Gnd**: This pin is connected to the ground terminal

The total time taken for the ultrasonic signals to echo back from an obstacle can be divided by 2 and then based on the speed of sound in air, the distance between the object and the HC-SR04 can be calculated.

We will see how to calculate the distance in the sketch for this device prototype. As per the HC-SR04 data sheet, it is a 5-volt tolerant device, operating at 15 mA, and has a measurement range starting from 2 centimeters to a maximum of 4 meters. The HC-SR04 can be directly connected to the Arduino board pins.

The 16x2 LCD character display is also a standard commercially available device, with 16 columns and 2 rows. The LCD is controlled by its 4 data pins/lines. We will also see how to send string outputs to the LCD from the Arduino sketch.

The power supply being used in today's example is a standard 9-volt battery plugged in to Arduino's DC IN Jack. Alternatively, another option is to use 6 pieces of either AA-sized or AAA-sized batteries in series and plug them into the VIN pin of the Arduino board.

Distance measurement device circuit

Follow the breadboard diagram shown next to build the distance measurement device. The diagram shown on the next page is quite complex. Take your time as you unravel through it. All the components (including the Arduino board) in this prototype are powered from the 9 volt battery.

Sometimes the LCD procured online might not ship with soldered header pins. In such a case, you will have to solder 16 header pins.

It is very important to note that unless the header pins are soldered properly into the LCD board, the LCD screen will not work correctly.

This is a very challenging prototype to get working in one go. Make sure there are no loose jumper wires.

Notice how the positive and negative terminals of the power source are plugged into the VIN and GND pins of the Arduino board respectively.

The 10K potentiometer has three legs. If you look straight at the breadboard diagram, the left hand side leg of the potentiometer is connected to the 5V power supply rail of the breadboard. Some typical potentiometers that can be used in this chapter are shown as follows for reference:

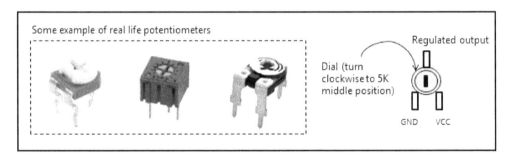

Figure 5: Typical potentiometers used in this chapter

The right hand-side leg is connected to the common ground rail of the breadboard. The leg in the middle is the regulated (via the potentiometer's 10K resistance dial) output that controls the LCD's V0/VEE pin. Basically, this pin controls the contrast of the display. You will also have to adjust (a simple screw driver may be used) the 10K potentiometer dial (to around halfway at 5K) to make the characters visible on the LCD screen. Initially, you may not see anything on the LCD, until the potentiometer is adjusted properly.

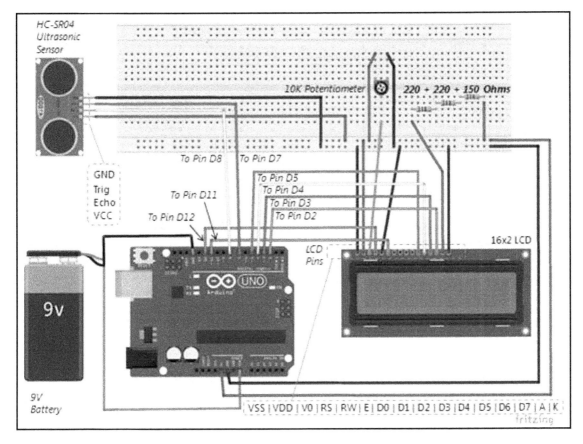

Figure 6: Distance measurement device prototype

When the 'Trig' pin receives a signal (via pin D8 in this example) this results it sending out ultrasonic waves to the surroundings. As soon as the ultrasonic waves collide with an obstacle, they get reflected. The reflected ultrasonic waves are received by the HC-SR04 sensor. The Echo pin is used to read the output of the ultrasonic sensor (via pin D7 in this example). The output read from the 'Echo' pin is processed by the Arduino sketch to calculate the distance.

The 16x2 LCD to Arduino Uno connections are tabulated next. Concise descriptions of the important pins are also provided here for easy understanding. For a detailed description, you may refer to the 16x2 LCD datasheet available online:

LCD pin	Arduino Uno pin	LCD pin description
VSS	GND (via breadboard)	Ground connection.
VDD/VCC	5V (via breadboard)	5V Power input.
V0/VEE	Potentiometer output	LCD contrast control receives output from potentiometer.
RS	Digital I/O Pin 12	Register select pin.
RW	GND (via breadboard)	Read/Write pin.
E	Digital I/O Pin 11	Enable pin.
D0	Not Used	These pins represent a 8-bit data. These pins are used to exchange data between the LCD and the Arduino board.
D1	Not Used	
D2	Not Used	
D3	Not Used	
D4	Digital I/O Pin 5	
D5	Digital I/O Pin 4	
D6	Digital I/O Pin 3	
D7	Digital I/O Pin 2	
A/LED+	5V (via breadboard and 590 Ohms resistors)	5V Power input
K/LED-	GND (via breadboard)	Ground connection

Table 4: LCD to Arduino Uno connections

The HC-SR04 ultrasonic sensor to Arduino Uno connections are tabulated next:

HC-SR04 Sensor pin	Arduino Uno pin
VCC	5V
Trig	Digital I/O Pin 8
Echo	Digital I/O Pin 7
GND	GND

Table 3: HC-SR04 to Arduino Uno connections

The power source to Arduino Uno connections are tabulated next:

Power source (9V)	Arduino Uno pin
Positive terminal	VIN
Negative terminal	GND

Table 5: Arduino power source connections

Distance measurement device sketch

Once you have assembled all the required parts together, load the following sketch. Before loading the sketch, make sure nothing is connected to Arduino's Tx (D1) and Rx (D0) pins.

The following sketch/code may be freely downloaded from the path mentioned in Chapter 1, *Boot Camp*. This is what a distance measurement sketch looks like.

```
//************************************************/
// Step-1: CONFIGURE VARIABLES
//************************************************/
#include <LiquidCrystal.h>
LiquidCrystallcd(12, 11, 5, 4, 3, 2);  // create LCD object
int trigPin = 8;
int echoPin = 7;

//************************************************/
// Step-2: INITIALIZE I/O PARAMETERS
//************************************************/
void setup()
{
  // one time setup code
  Serial.begin (9600);
```

```
  pinMode(trigPin, OUTPUT);
  pinMode(echoPin, INPUT);
  lcd.begin(16, 2);
  lcd.setCursor(0, 0);
  lcd.print("Distance:");
}

//**********************************************************/
// Step-3: MAIN PROGRAM
//**********************************************************/
void loop()
{
  long duration, distance;
  // Signal a quick LOW just before giving a HIGH signal
  digitalWrite(trigPin, LOW);
  delayMicroseconds(2);
  // After 2 micro-seconds of LOW signal, give a HIGH signal
  // to trigger the sensor
  digitalWrite(trigPin, HIGH);
  // Keep the digital signal HIGH for
  // at least 10 micro-seconds
  // (required by HC-SR04 to activate emission of
  // ultra-sonic waves
  delayMicroseconds(10);
  // After 10 micro-seconds, signal a LOW
   digitalWrite(trigPin, LOW);
  // Now wait for the Ultra sonic echo to return from an
  // obstacle
  duration = pulseIn(echoPin, HIGH);
  // Convert the distance to centimeters
  distance = (duration/2) / 29.1;
  // Print the distance on the Serial Monitor window
  Serial.print(distance);
  Serial.println(" cms");
  // Clear the LCD
  lcd.clear();
  // Set the cursor position
  lcd.setCursor(0, 0);
  // Print the distance on the LCD screen
  lcd.print("Distance (in cms):");
  lcd.setCursor(0, 1);
  lcd.print(distance);
  // Check for distance again
  delay(400);
}
```

Before operating the prototype, let's try to understand the sketch briefly:

The sketch starts with the customary section of including the Arduino header file for operating (or driving) the LCD panel using:

```
#include <LiquidCrystal.h>
```

Version 1.0.5 of the library has been used in this chapter.

This is followed by defining a global variable to refer to the LCD panel. Throughout the sketch, this `lcd` variable is used to interact with the LCD panel.

```
LiquidCrystallcd(12, 11, 5, 4, 3, 2);
```

By now, you will be able to understand the various lines of code in the `setup()` function. You will notice the familiar `pinMode()` function calls for configuring the pins interfacing with the ultrasonic sensor. The last few lines in the `setup()` function initialize the LCD panel.

```
lcd.begin(16, 2);
```

The next line sets the location of the cursor to column 0 row 0 in the 16x2 LCD array.

```
lcd.setCursor(0, 0);
```

The following line prints the `Distance` string on the LCD panel.

```
lcd.print("Distance:");
```

The following function call is used to read the duration taken by the ultrasonic waves to return to the HC-SR04 sensor.

```
duration = pulseIn(echoPin, HIGH);
```

The `pulseIn()` function is used to read and measure the duration of an input signal on a particular pin. Based on the logic level specified as a parameter to the function, it will wait for and read the signal until its logic level changes. In the preceding statement, when reading a HIGH signal, it waits for a HIGH input signal and measures the duration until the signal level changes to LOW. This is how the sketch measures the duration taken by the ultrasonic waves to return to the HC-SR04 sensor.

The next step is to process and convert the duration into centimeters using the following logic:

```
distance = (duration/2) / 29.1;
```

Eventually, clear the LCD screen.

```
lcd.clear();
```

As a last step, the sketch displays the distance on the LCD panel.

```
lcd.setCursor(0, 0);

lcd.print("Distance (in cms):");

lcd.setCursor(0, 1);

lcd.print(distance);
```

After the sketch has been loaded successfully, unplug the Arduino from the USB B port (make sure the Arduino board is no longer connected to the computer). The next step is to connect all the assembled components to the Arduino board using the breadboard.

Once all the parts have been connected, plug in the 9-volt battery power source into Arduino's VIN and GND pins (as shown in the breadboard setup diagram). As soon as the power source is plugged in, the device prototype should get powered up.

Operating the distance measurement device

Now, point the Ultrasonic sensor straight toward any object/obstacle (within 4 meters range) and the distance of the object will be displayed on the LCD screen. Slowly, move the entire device toward or away from the object/obstacle and observe how the distance displayed on the LCD screen changes.

Every time you move the device, wait for around 2 seconds in one position in order to get a stable reading on the LCD screen. Since the device will be handheld, the distance will keep changing due to mild movements of the human hand; therefore, to display a stable reading, the Arduino sketch has been written to run at every 2-second interval and display the calculated distance.

Finishing touches

We are almost done with our first standalone device prototype. Just two more things remain:

- A power switch
- A project enclosure

A simple power switch will be used to switch the power supply to the device ON and OFF. There are various types of power switches and numerous options to choose from. In order to make our selection easier, we should know that there are mainly two types of power switches:

- Momentary switches
- Maintained switch

Momentary switches are the common push button switches that we have seen so far. Recall the push button that we used in `Chapter 3`, *Day 1 - Building a Simple Prototype*. The main feature of a momentary switch is that it remains in a closed state (stays ON) as long as the button on the switch remains pressed (technically known as switch **Actuation**); at all other times, the momentary switch remains in an open state (remains OFF). Momentary switches are mostly used to switch on the power supply to a device that needs a momentary activation.

For the standalone device example that we learnt in this chapter, a **Maintained** switch would be an appropriate choice. As the name suggests, a maintained switch maintains the state (open or closed) unless the switch lever is physically moved (technically known as switch **Actuation**).

A maintained switch is very similar to normal wall switches that we use at home. The only difference is the size and the way the switches will be mounted with the Arduino project enclosures. As there are numerous types of switches available in the market, for simplicity, we will use a **Single Pole Single Throw** (**SPST**) toggle switch, shown in the following diagram:

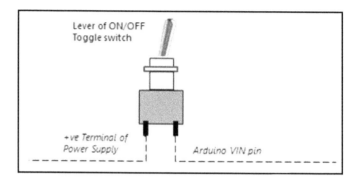

Figure 7: A typical ON/OFF toggle switch

All we have to do is to place the two terminals of the switch between the positive power supply terminal and the Arduino board, as shown in the following diagram:

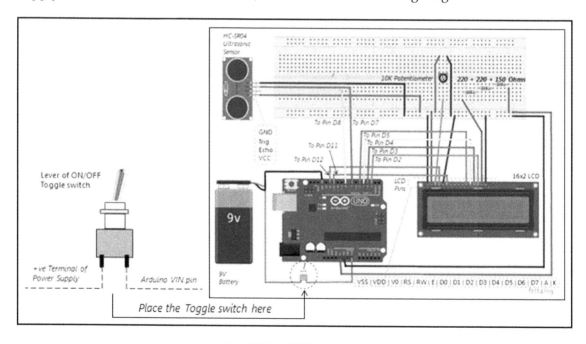

Figure 8: Placing a ON/OFF toggle switch

Occasionally, we might need to house the Arduino prototype in an enclosure or a box, especially during a presentation. There are various types of project enclosures that may be utilized for housing the device prototype. You can procure plastic-based transparent project enclosures from online stores. Some enclosures come specifically designed to house an Arduino, whereas, there are others that provide various other features such as battery holders, waterproof, dust proof casing, and so on.

Usually, for most prototyping needs, we will use simple plastic enclosures that are easy to procure. However, if by chance you want finesse, then you will find professional quality project enclosures at the following websites; choose wisely based on your needs:

- `https://www.amazon.com`
- `https://www.sparkfun.com`
- `https://www.adafruit.com`
- `https://www.itead.cc`
- `http://www.takachi-enclosure.com`

In recent times 3D printing has also emerged as a popular technique of creating innovative and customized project enclosures. This technique is primarily based on basic 3D printing technology. To think of it in a simple manner, basic 3D printing is done by systematically depositing specifically designed binding material vertically in layers until a three dimensional structure is created. To draw an analogy, if we draw a circle on a piece of paper it is an example of two dimensional or 2D printing. Whereas if we draw several circles on a piece of paper and then cut them out and individually stack them on top of one another in multiple layers then it would be analogous to 3D printing.

The main steps involved in creating 3D printed project boxes are:

- **Step-1**: Designing the 3D model with a professional 3D modelling CAD/CAM software. Alternately, for a easy start you can use tinkercad which is an online 3D designing and printing software you'll find at: `https://www.tinkercad.com/`.
- **Step-2**: Submit the 3D model file (usually known as STL or StereoLithography file) to an online 3D printing services vendor local to you.

The main advantage of 3D printed project boxes is that they are fully customized for your project and fit with a snap tight finish. But you will have to get used to the 3D printing process. You may either design the 3D models yourself and then submit the model to a commercial 3D printing vendor or you can search online repositories (such as `https://www.thingiverse.com`) for readymade project models and parts that can be downloaded and assembled in the 3D modelling software.

Try the following

Let's try the following things before proceeding to the next chapter:

- Try using a different pin for the echo. Make appropriate changes in the code and breadboard setup.
- Try displaying the distance in inches. Modify the calculation appropriately.
- Add a Red LED to the prototype and start blinking it as soon as something comes within 10 centimeters of the device.

Things to remember

Remember the following important points about what we learned in this chapter.

- Standalone devices operate from independent power sources
- Independent power sources can be in the form of batteries, wall AC to DC adapters and USB power hubs
- Remotely located devices may need to be powered from solar powered cells
- There are three ways to power the Arduino board from external sources; via USB, through the DC IN port, and by using the VIN pin
- 5-volt tolerant peripherals can be directly powered from the Arduino 5V power supply pin

Summary

In this chapter, we learned how to build a standalone device using the Arduino board as the controlling device. We saw how a standalone device can be easily built around the Arduino board by connecting multiple peripheral devices and powering it from independent power sources.

The HC-SR04 ultrasonic sensor that we used in this chapter can be used in a variety of real-world applications such as: automatic sliding doors, obstacle detectors for toys/robots, depth/distance measurement, and so on. Similarly, you can now start imagining numerous standalone devices that can be built around the Arduino board.

In the next chapter, we will learn how to use Actuators (components that move such as DC motors, Servo motors, and so on). If you look around, there are numerous examples of actuators such as the application of motors in gadgets with moveable parts. Actuators are usually powered from independent power sources to avoid damage to the Arduino board due to the high power requirements of actuators.

7

Day 5 - Using Actuators

"Give me a lever long enough and a fulcrum on which to place it; and I shall move the world."
- Archimedes

As promised in `Chapter 3`, *Day 1 - Building a Simple Prototype*, today we will work on our first project that uses diodes and transistors with a DC motor (an example of an actuator) powered from an independent battery-based power source. This is an advanced level chapter and is designed with a lot of concepts and components; it builds upon the knowledge gathered so far during the previous chapters. In this chapter, our focus will be on moving parts that are used to physically move things.

You will learn the following topics in this chapter:

- Interfacing a DC motor
- Powering DC motors from a battery
- Using multiple power sources
- Concept of Common Grounding
- Concept of reverse current
- Practical use of a diode
- Practical usage of a transistor
- Concept of pull up/pull down resistors
- Concept of PWM
- DC motor speed control
- Using interrupts for I/O processing
- Interfacing a servo motor

About actuators

An actuator is an electro-mechanical device that translates electrical energy into motion. A **DC motor** is a perfect example of a **basic actuator**. Similarly, servo motors, stepper motors, and hydraulic arm levers are all examples of actuators that are used heavily in the world of hardware automation and robotics.

DC motors are of two main types: brushed and brushless. There are fundamental differences in the way brushed and brushless motors are designed. To think of it in a very simple manner, brushed DC motors make use of physical contact points, known as brushes, between their current supply and the Commutator (the motor part responsible for causing the rotor to move). Whereas, brushless DC motors do not use physical brushes; instead they usually use multiple permanent magnets around the rotor, but not in direct contact. Brushed motors are the commonly found low cost, noisy motors that we are used to seeing in toy cars. Whereas, the brushless motors are more sophisticated, almost noiseless and higher priced components; usually found inside expensive electronic gadgets such as camcorder shutters. In this chapter, we will take a look at the basic aspects of interfacing and controlling easily available brushed DC motors.

Special considerations while using DC motors

Very small DC motors that have current requirements of less than 30 mA may be run for very short periods (a few seconds) of time by directly powering them from the Arduino pins. This may be done for experimental purposes only; however, in practical settings, directly powering a DC motor from the Arduino pins is not recommended.

Unlike other DC powered components that we have seen so far, the average DC motor requires a lot of current, much more than what can be supplied by an Arduino Uno pin. A separate power source must be used for powering DC motors. Hence grasping the concept of using multiple power sources in a single standalone circuit becomes extremely essential.

Fundamental - Multiple power sources:
When working with devices that require large currents that cannot be supplied by the Arduino board, separate power sources should be used. One for powering the Arduino (the brain of the circuit) and another more powerful source for the DC motor (the brawn of the circuit). This technique ensures separation of the power sources: one for logic and the other for movement.

Before writing the C sketch for the Arduino, there is one more basic aspect of a DC motor that must be understood. A reverse current may be generated when the DC motor stops.

The reverse current has the potential of flowing back into other electronic components (including the Arduino pins) in the circuit - this might lead to the damage of delicate electronic parts (including an Arduino pin).

In order to stop this reverse current from flowing back, a diode is placed in an appropriate manner in the circuit. Rule of thumb is to place the striped end of the diode towards the positive terminal of the power source powering the motor. We will see how this is achieved shortly in the following sections:

You must learn to place the push button correctly on the breadboard; if it is placed incorrectly then the button state will always be HIGH, therefore resulting in the DC motor running even without pressing the button. Refer to `Chapter 3`, *Day 1- Building a Simple Prototype* for the fundamentals of using a push button.

In order to give the reader a rounded experience, in the first section, there will be an example of running a very small DC motor by powering it from an external battery power source. The example will demonstrate how the motor can be operated using a push button.

In the second section, the concept of Pulse Width Modulation for controlling the speed of a DC motor will be introduced. In this example, we will look at powering the DC motor from a separate external power source and controlling its running speed using multiple momentary push buttons.

A basic DC motor prototype

In this section, we will build a prototype where a DC motor will spin for a second every time a push button is pressed.

For building the distance measurement device, the following parts will be required:

- Arduino Uno R3
- USB connector
- Four pieces 1.5V batteries
- One battery holder (for 4 batteries)
- One full sized breadboard
- One push button
- One piece 10K Ohms resistor

- One small DC motor
- One N2222 transistor
- One IN4001/IN4007 diode
- Two pieces 150 Ohms resistors
- Some Jumper wires

Follow the schematic diagram shown below to build the DC motor circuit. In the following breadboard diagram, notice that an external battery-based power source has been used for powering the DC motor; whereas the Arduino board has been powered via the computer USB.

Using a separate power source for the DC motor is important, because the 5V pin of the Arduino board is not designed to provide the amount of current drawn by a DC motor for continuous usage. The next diagram depicts how to setup the breadboard circuit for our first DC motor prototype.

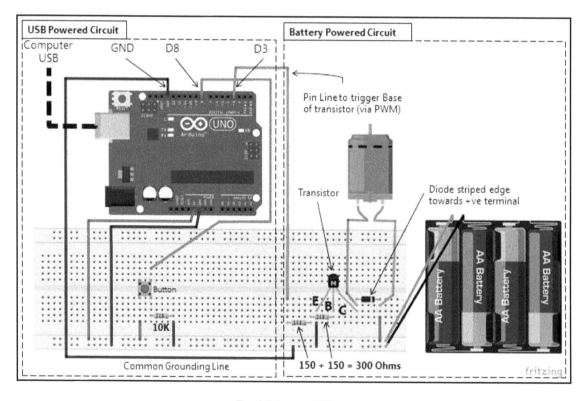

Figure 1: Button operated DC motor

Note how the breadboard has been used to keep the two power sources separate from one another.

The power rails of a full-sized breadboard are physically divided into a left and a right side circuit internally--each side having five sections. Make sure you understand how to use the breadboard. Just think of a full sized breadboard as two half sized breadboards, placed side by side together and it will become easy to understand.

In the preceding diagram, the push buttons are powered by the Arduino using the five sections on the left side of the breadboard power rails, while the DC motor circuit is powered from the battery by using the five sections on the right-hand side.

Fundamental - Common Grounding:
Also notice how the GND pin from the Arduino has been connected to the DC motor circuit on the right side of the breadboard. This has been done to achieve common grounding in a circuit that uses multiple power sources.

As shown in Figure 1, first the Arduino is powered from the computer USB, the Arduino in turn provides power to the push button via its five-volt pin. The second power source is provided from the battery; here four standard 1.5 volt batteries have been used. The battery is used to power the DC motor.

Caution:
Do not use the Arduino board pins to power the DC motor directly as the Arduino board and/or power supply pins might get damaged due to direct prolonged usage with a DC motor.

The example in this chapter should be used for prototyping and short-term usage only. Ideally, a motor driver circuit should be used for prolonged use. A professional motor driver circuit comes with in-built protection so that the Arduino or the micro-controller used to drive the motor does not get damaged.

Additionally, it must be noted that digital pin 3 from Arduino has been connected to the base of the transistor. When a PWM signal is received via pin 3 on the transistor base, the motor circuit gets closed and the motor starts rotating.

You might have noticed that a push button has been used with the Arduino board in the preceding circuit diagram. Note the three connections (5V, GND, and digital pin) for a push button. The push button in this example has been used as a pull-down resistor. It is a resistor used to connect an electronic signal point/pin to the negative terminal or GND of the circuit's power supply. Doing so keeps the logic level at LOW when the button is not pressed. As soon as the button gets pressed, the logic level changes to HIGH.

Similarly, a pull-up resistor is used to connect an electronic signal point/pin to the positive terminal of a power supply in circuit. Doing so keeps the logic level at HIGH when the button is not pressed. As soon as the button is pressed, the logic level changes to LOW.

 The button is grounded (LOW) via 100 Ohms pull-down (because it connects to the ground pin) resistor. This technique is used to ensure that when the statement `digitalRead(buttonPin)` is executed, it will return a LOW (if the button is NOT pressed) and will return HIGH (if the button IS pressed).

Basic DC motor sketch

The C sketch for running the DC motor using a push button is provided in the following. The sketch is designed to run the motor for 1 second and then stop it every time the push button is pressed.

 The code in this chapter may be freely downloaded from the location for this chapter mentioned in Chapter 1, *Boot Camp*.

The following code is a basic DC motor sketch:

```
//***********************************************************/
// Step-1: CONFIGURE VARIABLES
//***********************************************************/
int motorPin = 3;        //this is a PWM capable pin
int buttonPin = 8;
int buttonState = LOW;

//***********************************************************/
// Step-2: INITIALIZE I/O PARAMETERS
//***********************************************************/
void setup()
{
  pinMode(motorPin, OUTPUT);
```

```
  pinMode(buttonPin, INPUT);
}


//**********************************************************/
// Step-3: MAIN PROGRAM
//**********************************************************/
void loop()
{
  buttonState = digitalRead(buttonPin);

  if(buttonState == HIGH)
  {
    analogWrite(motorPin, 160);    //run the motor
    delay(1000);
    analogWrite(motorPin, 0);    //stop the motor
  }
}
```

Now let us understand the preceding program. As soon as the push button is pressed, the program reads the status via pin 2 and enters the if block and executes the `analogWrite(motorPin, HIGH)` statement; sending an analog signal on pin number 3, which in turn switches the transistor ON. This allows the current to flow across the motor and make it run.

After a delay of 1 second, the `analogWrite(motorPin, 0)` statement executes and sends a near zero voltage signal on pin 3. The very low signal turns the transistor OFF, thereby breaking off the power supply circuit of the DC motor. The preceding cycle is executed every time the push button is pressed.

DC motor speed control - PWM method

In this section, an important fundamental known as **Pulse Width Modulation (PWM)** will be introduced. So far in the examples, the `analogRead()` and `analogWrite()` functions were used with the analog I/O pins. However, for achieving PWM, the new thing that we will do is to use the `analogWrite()` function to send signals on digital pin 3. We will only change the C sketch slightly and retain the same circuit that was setup during the previous example; but before that let us quickly understand PWM.

A signal (or pulse) is basically represented by a voltage level on a particular pin spanning for a certain amount of time (known as the width of the signal/pulse). Being an advanced topic, only the bare minimum required fundamentals will be discussed here. The speed of a motor is controlled by regulating the input voltage to the motor. Conventionally, impedance (resistance) is used to regulate the input voltage and control the speed of motors.

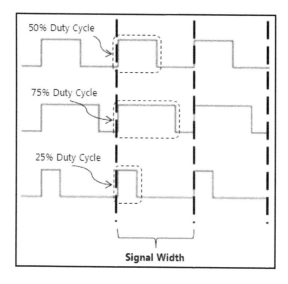

Figure 2: Pulse Width Modulation by varying duty cycles

The PWM method is used as an electronic mechanism to control the speed of DC motors without using voltage reducing methods. In the PWM method, the duration (pulse width) of a pulse during which the voltage remains HIGH is controlled (modulation).

Using the `analogWrite(pwm-pin, parameter)` function, a number between 0 and 255 can be passed as a parameter. When `analogWrite(pwm-pin, 255)` is executed, the signal voltage level remains HIGH for the entire duration (width) of the pulse, that is a 100% of the pulse width, also known as 100% duty cycle.

When `analogWrite(pin-number, 0)` is executed, the signal voltage level remains LOW for the entire duration of the signal (width), that is 0% of the pulse width, also known as a 0% duty cycle. When `analogWrite(pin-number, 128)` is executed, the signal voltage level remains HIGH for half of the duration of the signal (or pulse),that is 50% of the pulse width, also known as a 50% duty cycle. So, when the DC motor power is using PWM it means that the power is being constantly switched ON and OFF, as depicted in the preceding figure.

Imagine what will happen if an electric fan is manually switched ON and OFF. If done slowly then the start and stop of the fan will be noticeable and it will run slowly; however, if the switching is done very fast then the start and stop of the fan will become less noticeable and it will rotate with a much higher speed. Similarly, the PWM pins on the Arduino Uno boards are designed to provide a way to run digital signals at specified duty cycles.

Fundamental: Signal Duty Cycle:
The percentage of a signal period (pulse width) during which the signal is active or in a HIGH state is referred to as the signal duty cycle.

There are six digital I/O pins on the Arduino Uno board that are marked with a tilde (~) symbol. These I/O pins are internally wired to provide PWM.

There are some known issues and variances on different PWM pins of the Arduino. There are also some known issues that affect the pulse width and hence all Arduino pins do not have the same frequency. Hence without going into these complications, think of varying the duty cycle as a number between 0 and 255, with zero being the lowest speed (ideally OFF position) and 255 being the highest speed. The speed of the motor will be directly proportional to the duty cycle of the pulse.

DC motor speed control sketch

For learning motor speed control, the same breadboard setup from the first part of this chapter may be reused as is - simply add two more push buttons to the breadboard, as shown below. Remember that the button connected to digital I/O Pin 8 corresponds to **Button 1** (the left most button on the breadboard) labelled in the following circuit:

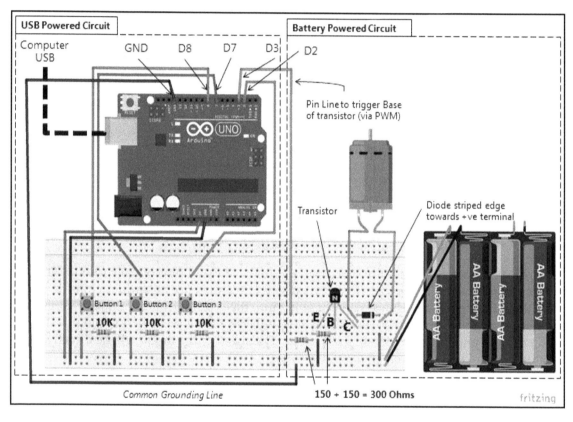

Figure 3: DC motor speed control prototype

The preceding diagram should have become self-explanatory by now. Notice the two separate power sources, the common grounding technique and the concept of using a transistor (explained in the chapter for Day 2) and diode (explained in the chapter for Day 2). Build the circuit, load the sketch provided later in this chapter, and start running the prototype.

The following C sketch is designed to run the motor at full speed for 1 second and then drop the speed to 75% and run the motor at 75% speed for another 1 second, every time a push button is pressed.

Here is the DC motor speed control code:

```
//*********************************************************/
// Step-1: CONFIGURE VARIABLES
//*********************************************************/
int motorPin = 3;        //this is a PWM capable pin
int buttonPin = 8;       //button to start at low speed
int buttonPin2 = 7;      //button to start at high speed
int buttonPin3 = 2;      //button to stop
int buttonState = LOW;
int buttonState2 = LOW;
int buttonState3 = LOW;
//*********************************************************/
// Step-2: INITIALIZE I/O PARAMETERS
//*********************************************************/
void setup()
{
  Serial.begin(9600);
  pinMode(motorPin, OUTPUT);
  pinMode(buttonPin, INPUT);
  pinMode(buttonPin2, INPUT);
  pinMode(buttonPin3, INPUT);
}
//*********************************************************/
// Step-3: MAIN PROGRAM
//*********************************************************/
void loop()
{
  buttonState = digitalRead(buttonPin);
  buttonState2 = digitalRead(buttonPin2);
  buttonState3 = digitalRead(buttonPin3);

  if(buttonState == HIGH)
  {
    Serial.println("button 1");
    delay(1000);
    analogWrite(motorPin, 170);   //run at low speed
  }
  else if(buttonState2 == HIGH)
  {
    Serial.println("button 2");
    delay(1000);
    analogWrite(motorPin, 240);   //run at high speed
  }
```

```
    else if(buttonState3 == HIGH)
    {
      Serial.println("button 3");
      delay(1000);
      analogWrite(motorPin, 0);    //stop the motor
    }
  }
```

Let us understand the preceding program, step by step. As soon as the push button is pressed, the program reads the status via pin 2 and enters the if block and executes the `analogWrite(motorPin, 255)` line; sending a signal with a duty cycle of 100% on pin number 3, which in turn switches the transistor ON (for 100% of the time). This allows the current to flow across the motor and make it run at full speed.

When the program flow executes the `analogWrite(motorPin, 192)` line, it sends a signal with a duty cycle of 75% on pin number 3, which in turn switches the transistor ON (for 75% of the time). This allows the current to flow across the motor and make it run at 75% speed. This cycle is executed every time the button is pressed.

Using Arduino interrupts

So far, we have seen how to use the push buttons to change the speed of the DC motor. The previous sketch is written to continuously loop and check for the state of each push button one by one in a sequential manner. First the sketch checks button 1, followed by button 2, and then button 3 - this sequence of checking the state of the buttons continues endlessly. This method of checking for user input may lead to timing issues. Let us understand how.

Let us say you press button 3 (to stop the motor) and at that point in time, if the sketch was checking the state of button 1, then we would have to wait for the sketch to first finish checking button 1, then button 2, and finally button 3 would be checked. Since our sketch is small in size and not doing too many things, you will hardly notice any issues. However, in real-world situations, this is not an ideal way to handle time-sensitive input events (for example, a button press).

Time sensitive input events should be handled in a responsive manner. In order to achieve responsiveness, Arduino provides the mechanism of interrupts. Interrupts provide a real-time reaction (signal processing) mechanism in the micro-controller world. It provides the ability to execute a function as soon as there is a change in the signal level on a pin (connected to a peripheral device like a simple button in the next example).

 Only digital pins 2 and 3 can be used for the purpose of interrupts in the Arduino Uno board. You can refer to the following URL for a comprehensive list of Arduino boards and their interrupt pins: https://www.arduino.cc/en/Reference/AttachInterrupt

To demonstrate the interrupt handling technique, we will utilize the previous example with three push buttons controlling the DC motor speed. In this case, we will make a very small change to the circuit setup and the C sketch.

In the C sketch, we are going to designate digital pin 2 (for receiving the stop signal via button 3) as an interrupt and associate this pin to execute a pre-defined C function as soon as button 3 is pressed. After loading and executing this sketch, you may notice a slight difference in the responsiveness of button 3 (for stopping the motor), as the sketch will directly jump to executing the function attached with the pin for button 3. In real-world situations, this slight improvement in responsiveness can work wonders and solve many timing and peripheral device synchronization issues.

Go ahead and load the following sketch for learning Arduino Interrupts into your board:

```
//**********************************************************/
// Step-1: CONFIGURE VARIABLES
//**********************************************************/
int motorPin = 3;        //this is a PWM capable pin
int buttonPin = 8;       //button to start at low speed
int buttonPin2 = 7;      //button to start at high speed
int buttonPin3 = 2;      //button to stop
int buttonState = LOW;
int buttonState2 = LOW;
int buttonState3 = LOW;
//**********************************************************/
// Step-2: INITIALIZE I/O PARAMETERS
//**********************************************************/
void setup()
{
  Serial.begin(9600);
  pinMode(motorPin, OUTPUT);
  pinMode(buttonPin, INPUT);
  pinMode(buttonPin2, INPUT);
  pinMode(buttonPin3, INPUT);
```

```
  // This line specifies that as soon as Pin 2 (for Button 3)
  // is HIGH, the stop() function should be executed
  attachInterrupt(digitalPinToInterrupt(buttonPin3),stop,HIGH);
}
//*********************************************************/
// Step-3: MAIN PROGRAM
//*********************************************************/
void loop()
{
  buttonState = digitalRead(buttonPin);
  buttonState2 = digitalRead(buttonPin2);
  buttonState3 = digitalRead(buttonPin3);

  if(buttonState == HIGH)
  {
    analogWrite(motorPin, 170);    //run at low speed
  }
  else if(buttonState2 == HIGH)
  {
    analogWrite(motorPin, 240);    //run at high speed
  }
}

void stop()
{
  analogWrite(motorPin, 0);    //stop the motor
}
```

Interfacing with a servo motor

Let us take a look at yet another very useful example of an actuator: the servo motor. A servo motor is a special type of motor that is capable of rotating its shaft at specified angles between 0 to 180 degrees.

A DC powered servo motor has numerous applications in electronic gadgets such as surveillance cameras, DVD players, and basically any application where things need to be moved at particular angles. AC powered servo motors are usually applied in the manufacturing industry to move machinery parts at specified angles - a common example would be a bottling plant:

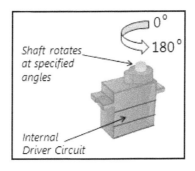

Figure 4: A typical servo motor

Internally, a servo motor has a regular DC motor that is integrated with an in-built motor driver circuit. The in-built driver circuit takes care of moving the motor at specified angles. Signals are sent from the Arduino to the servo motor, via an Arduino library. The Arduino library that we will use in this example is `Servo.h`.

In the following example, we will learn how to interface and use a DC powered servo motor with the Arduino. We will need the following components for setting up the Servo motor circuit with the Arduino board:

- Arduino Uno R3
- USB connector
- One full sized breadboard
- Two push buttons
- Two pieces 10K Ohms resistor
- One SG90 servo motor
- Some jumper wires

Servo motor control circuit

The Servo motor circuit setup is very straightforward. The breadboard setup of the servo motor with Arduino Uno is shown in the following figure for reference. Follow the schematic diagram shown below to build the DC motor circuit.

This setup has been designed in such a way that every time upon pressing the right-hand side button, the servo motor's rotor position will get incremented by 5 degrees, until it reaches 175 degrees. Similarly, upon pressing the left-hand side button, the position of the servo motor's rotor will get decremented by 5 degrees:

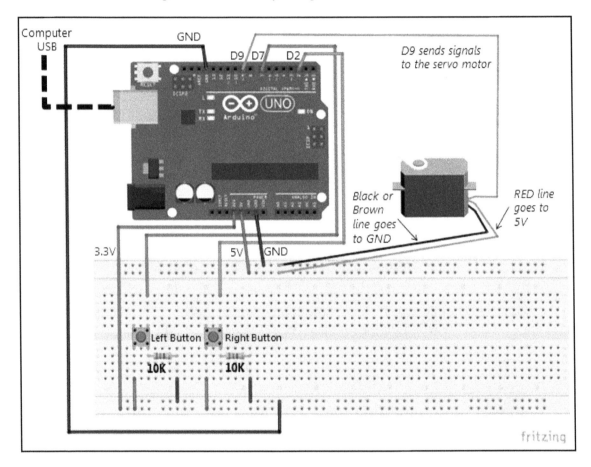

Figure 5: Servo motor breadboard setup

Note how the two momentary push buttons have been powered from the 3.3V pin. Since the Arduino Uno treats 3.3 volt signals as HIGH, we can use the 3.3V pin with the push buttons and read the HIGH signals using the `digitalRead()` function.

Start running this prototype by pressing the right-hand side button as the right side button press is coded to increment the servo motor's rotor position from zero. If you try to press the left side button in the beginning, then there will be no response, as the servo motor position will be set to zero degrees to begin with.

Servo motor control sketch

After building the circuit, load the following sketch into your Arduino Uno board. This sketch for controlling a servo motor may also be downloaded from the online location for this chapter mentioned in Chapter 1, *Boot Camp* of this book.

```
#include <Servo.h>// Arduino header file for servo motor

Servo servo;            // Servo motor object

int leftButton = 7;     // pin for reading left button state
int servoPin = 9;       // pin for sending signals to the servo motor
int rightButton = 2;    // pin for reading right button state

int angle = 0;                  // to track servo position in degrees
bool leftPressed = false;    // flag to track pressed button
bool rightPressed = false;   // flag to track pressed button

void setup()
{
  Serial.begin(9600);
  servo.attach(servoPin);     // specify the pin for controlling
                              // the Servo motor
}

void loop()
{
  // do not press both buttons at once
  // otherwise the sketch logic will get confused
  leftPressed = digitalRead(leftButton);
  rightPressed = digitalRead(rightButton);

  if(leftPressed)
  {
    decreaseAngle();
    Serial.println("left");
    Serial.println(angle);
  }
```

```
    if(rightPressed)
    {
      increaseAngle();
      Serial.println("right");
      Serial.println(angle);
    }
}

void decreaseAngle()
{
  // decrease angle
  if (angle > 10)
  {
    angle = angle - 5;
    servo.write(angle);
    delay(1000);
  }
}

void increaseAngle()
{
  // increase angle
  if (angle < 170)
  {
    angle = angle + 5;
    servo.write(angle);
    delay(1000);
  }
}
```

Now let us try to understand how the sketch is interacting with the servo motor. To start with, we have to use the Arduino header file for interfacing with the servo motor. This header file contains all the important functionality that is required for controlling a servo motor:

```
#include <Servo.h>
```

Thereafter, declare all the variables required for this sketch. The following object has to be declared in order to obtain a reference to the servo motor that we will be controlling:

```
Servo servo;
```

The other variables are for the three digital pins used to read the state of the momentary push buttons and to send signals to the servo motor:

```
int leftButton = 7;
int servoPin = 9;
int rightButton = 2;
```

At the beginning of the sketch execution life cycle, we will have to specify or attach the digital pin that will be used by the functionality within the servo header file. Throughout the execution of this sketch, this particular pin will be used to send signals to the servo motor:

```
servo.attach(servoPin);
```

The sketch executes in a continuous loop. At the beginning of the `loop()` function, the program logic samples the logical state of the two push buttons:

```
leftPressed = digitalRead(leftButton);
rightPressed = digitalRead(rightButton);
```

Depending upon which of the two buttons have been pressed, the servo position is changed. If the right-hand side button is pressed, then the `increaseAngle()` function is invoked. The logic in this function increases the value of the variable used to track the most recent angle of the servo motor:

```
angle = angle + 5;
```

If the left-hand side button is pressed, then the `decreaseAngle()` function is invoked. The logic in this function decreases the value of the variable used to track the most recent angle of the servo motor:

```
angle = angle - 5;
```

The last step is to change the servo motor's angle by sending the value of the `angle` variable to the servo via the following function call:

```
servo.write(angle);
```

Thus, we conclude our study of using a servo motor with the Arduino Uno.

Future inspiration

You can now apply this information to build innovative moving devices such as a remote-controlled reconnaissance boat that can be guided into dark, underground drainage systems in smart cities. The primary focus of this section will be to inspire you by giving you an advanced idea that you should try out on your own.

Sounds exciting, right? Take a look at the diagram below. You will get an idea of how we can build a remote-controlled boat using a DC motor and a servo motor. You would also need additional devices such as a radio frequency module for operating the motors remotely. We will learn Radio Frequency-based communications in `Chapter 10`, *Day 8 - Short Range Wireless Communications*.

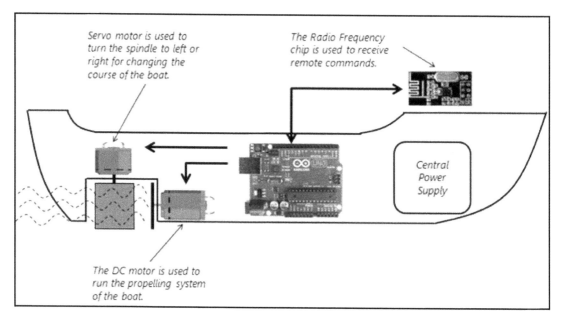

Figure 6: A wireless drone boat

The inspirational preceding example uses one DC motor and a servo motor. The DC motor has to be fitted to drive the boat's propulsion system, while the servo motor has to be fitted and positioned appropriately to turn the boat's navigation tail. The boat will be mounted with an Arduino Uno and a Radio Frequency chip.

The RF chip will listen for and receive RF-based commands sent to the boat unit from a remote location. Based on the command sent, either of the following devices will be operated:

- The DC motor, to propel the boat forward
- The Servo motor to turn the boat to the left or to the right

Try the following

Let us try the following points before moving to the next chapter:

- Try adding a fourth button in the speed control setup. This fourth button should run the motor by using the function call `analogWrite(motorPin, 200)`.
- In the servo motor example, change the 5 degrees increment and decrement to 10 degrees. Notice how the rotation speed increases with an increase in the change in angle.

Things to remember

Remember the following points before moving to the next chapter.

- Do not power motors directly from the Arduino board
- Arduino's GND pin is used to achieve a common ground between multiple power sources
- The transistor is continuously switched ON and OFF by using the `analogWrite()` function to send PWM signals via a PWM capable digital pin
- The `analogWrite()` function can be used with Arduino's digital pins
- Interrupts are used to handle user inputs in a responsive manner

Summary

In this chapter, we learnt how to work with actuators. We started working with a DC motor; then we learnt how to control its speed using the PWM method. Next, we learnt how to use Arduino interrupts and why they are important. In the end, we worked with a servo motor and understood how to move a servo motor at specified angles.

As a conclusion, we looked at a suggested drone-boat design using some motors and a radio frequency module with the Arduino. In this chapter, we already learnt how to use a DC motor and a servo motor. After reading through `Chapter 10`, *Day 8 - Short Range Wireless Communications* you will be able to use a radio frequency module with the Arduino. Thus, we come to the end of our short journey of learning to use actuators.

In the next chapter, we will learn how to interface AC powered devices with the Arduino. AC mains are at a very high voltage and current rating compared to Arduino's DC power rating. Therefore, we will learn the fundamentals of interfacing AC powered devices with the Arduino. This chapter will enable us to apply our knowledge for home automation projects.

8

Day 6 - Using AC Powered Components

"If you want to find the secrets of the universe, think in terms of energy, frequency and vibration."
- Nikola Tesla

So far, we have been working with DC powered devices. This chapter of the crash-course book will introduce us to the fundamentals of interfacing and controlling AC powered electrical devices with the Arduino platform.

This topic was specifically chosen for a wholesome completeness of the ten-day crash-course challenge that the book offered and you bravely undertook. We have surely come a long way since blinking that LED on the first day!

You will learn the following topics in this chapter:

- The concept of separating a DC circuit and AC circuit by using a relay module
- Understanding parts of a relay module
- Interfacing a relay module with the Arduino
- Interfacing a sound detector sensor with Arduino

Using relays with AC powered devices

In this section, we will learn the concept of separating a DC circuit from an AC circuit by using a relay device. AC powered devices operate at dangerous voltages and currents, whereas the Arduino Uno is a 5V, low current device in comparison to an AC powered device. Therefore, in order to interface the Arduino with an AC powered device, an intermediate device known as a relay is used. A typical relay device is shown in the following figure:

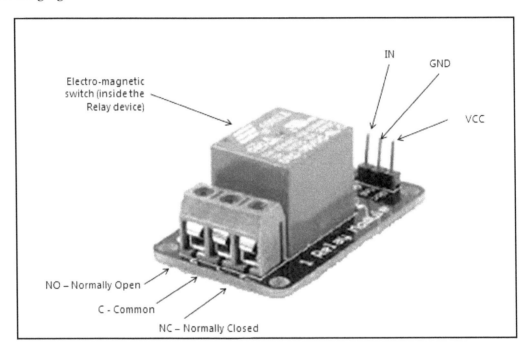

Figure 1: A typical relay device

As we can see, there are three socket type ports on the left-hand side of the Relay device. Usually, these ports are marked with abbreviations such as: **NO, C,** and **NC**. The usage of these three sockets is explained as follows:

- **Normally Open (NO)**: This port is called 'Normally Open' because this port is always disconnected (Open) from the AC circuit, via the common terminal.
- **Common (C)**: Think of this port as a common connection line between the NO and the NC ports. One end of the power supply terminal should be connected to the Common port.

- **Normally Closed (NC)**: This port is called 'Normally Closed' because this port is always connected (Closed) to the AC circuit, via the Common terminal.

A relay is an electrically (sometimes electro-magnetically) operated switch. A relay can be safely switched ON by a small current (available from an Arduino pin) and in turn the relay switches ON a circuit operating at a much larger current and voltage (AC powered circuits and devices):

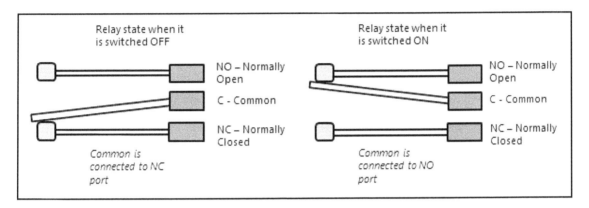

Figure 2: Logical representation of the internal working of a simple relay

For a better understanding of how the Relay module works internally, refer to the preceding diagram. The preceding diagram depicts a logical connection between the three ports on a Relay device. Note that the above diagram has been created for explaining the internal logic only, it is not the official design of a relay device.

When the Relay is switched off, then the Common port is internally disconnected from the NO port, as shown on the left hand side of the above figure. So if an electrical device is connected to the NO port then the electrical device will remain in a switched OFF state.

However, when the Relay is switched on (triggered by a small input voltage), then the Common port gets internally connected to the NO port, as depicted on the right-hand side of the preceding diagram. So if an electrical device is connected to the NO port then the electrical device will get switched ON.

In this example, we will see how to interface a relay with the Arduino board and operate (switch ON and OFF) a light bulb. Additionally, we will also learn how to interface a sound detector sensor device with the Arduino board. We will build a prototype to switch a light bulb ON and OFF by simply snapping our fingers or clapping our hands. Sounds like magic!

Fundamental:
The main reason for using a relay is to safely operate a higher power device from a lower power device. The relay separates the DC circuit from the AC circuit.

Typically, the following ports/jacks in a relay are used for making wiring connections. The following ports/jacks are tabulated for reference:

Port/Jack	Arduino Uno input port	Purpose
COM	Common connection	Terminal for connecting the AC power supply line to an AC device.
NC	Normally Closed connection	Always closed means always in contact with the COM, even when the relay is not powered up.
NO	Normally Open connection	Always open means in a state of being disconnected from the COM port/jack.
IN 1	Input pin 1	For receiving triggering logic from an Arduino pin.

Table 1: Connection points of a relay module

In the following sections, we will see how a simple AC powered device such as a light bulb can be interfaced and operated using the Arduino platform. Since AC mains power supply is involved, some of us might not feel very comfortable about handling this chapter. Therefore, this chapter has been divided into two practical hands-on sections.

The first section will provide a hands-on lesson for simulating a situation similar to an AC powered device using a relay with the Arduino. Instead of using AC mains, we will use a DC power supply, while a red LED will be used in place of an electric bulb.

If you are not comfortable dealing with AC power, then attempt the second part of this chapter only when you are confident about working with AC mains. The basic fundamentals of using a relay (with AC power) will get covered in the first part of the chapter anyway.

Even if you do not try the second part immediately, you will not be missing out on anything as the fundamental approach of using a relay device for isolating AC powered circuits from a DC powered circuit would have been already covered in the first part of this chapter. The C sketch in this chapter can be used for both part one and part two.

Part 1 - Simulation of sound activated light bulb controller

In this section, we will simulate the usage of a relay device with an LED. Once we understand the basics of interfacing a relay device, we will move on to using an actual AC bulb with a relay.

For building the simulated prototype the following parts will be required:

- Arduino Uno R3
- USB connector
- Two pieces 1.5V AA batteries + holder case
- One sound detector sensor module
- One 5V 10A AC power relay
- One red LED
- One 100 Ohms resistor
- Jumper wires as required

Purchase the relay based on the AC mains rating in your country. For example, in some countries the voltage rating is 220V, whereas in other countries it may be 110V.

Before moving on, since we will also use a sound detector module with our prototype, let us quickly take a look at the basic pin out of a typical sound detector device:

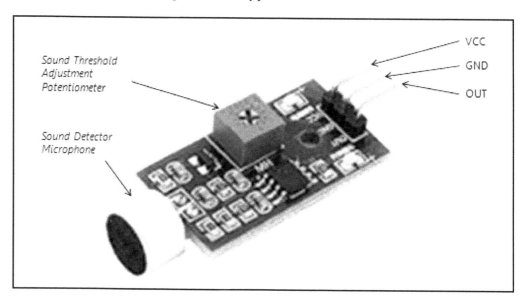

Figure 3: Sound detector module

As we can see in the preceding figure, a typical sound detector module has the following main parts:

- **Microphone:** The microphone is used to receive ambient sound waves.
- **Potentiometer:** The potentiometer allows us to manually adjust the sound level threshold at which the sound detector module will detect ambient sound. Depending upon the sound detector module, you will have to perform some trial and error for finding out the right position at which the potentiometer must be kept at, for meeting your project requirements. For the sketch in this chapter, there is no need to manually modify the position of the potentiometer.

Once all the components are available, follow the breadboard diagram shown to build the circuit:

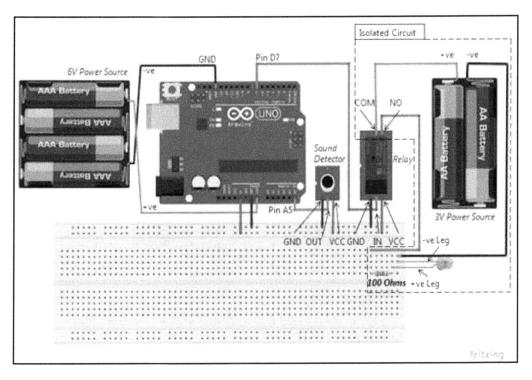

Figure 4: Sound-activated light bulb simulation with LED and DC power supply

As shown in the preceding figure, the sound detector to Arduino Uno connections are tabulated as follows:

Arduino Uno pin	Sound sensor pin
Analog I/O pin A5	OUT
5V	VCC
GND	GND

Table 2: Sound sensor to Arduino connections

As shown in the preceding figure, the relay to Arduino Uno connections are tabulated as follows:

Arduino Uno pin	Relay Port/Jack/Pin
Digital I/O pin 7	IN
5V	VCC
GND	GND

Table 3: Relay to Arduino connections

The relay to DC power and LED connections are tabulated as follows:

Relay port	DC Power/Device
COM	Positive terminal
NO	To LED (via bread board)

Table 4: DC devices to relay connections

The positive terminal of the power supply should be connected to the COM terminal of the relay device. The negative terminal of the power supply should be connected to the other end of the LED. Once everything is in place, load the sketch mentioned in a later part of this chapter and see how the sound detector triggers the switching of the LED light. All you need to do is to snap your fingers in front of the sound detector.

The sound-activated device sketch

The C sketch is designed to switch ON the light as soon as there is a sound detected by the sound sensor. If another subsequent sound is detected, the light will be switched OFF. This sketch can be used with both parts in this chapter.

By now, you should be able to understand the following code on your own. However, we will take a quick look at the various portions of the code shortly.

The sketch is available for download from the online location mentioned in `Chapter 1`, *Boot Camp* of this book.

Here is the sketch for a sound operated relay switch:

```
//**********************************************************/
// Step-1: CONFIGURE VARIABLES
//**********************************************************/
int _pinD10 = 10;                  // pin 10 to receive input
                                   // from sensor
int _pinD7 = 7;                    // pin 7 to send signal to
                                   // relay
int _soundLevel = HIGH;            // variable for storing
                                   // output of sensor
boolean _bRelayState = false;      // start by assuming OFF
                                   // state
unsigned long _lastSoundEventTime; // variable to store the
                                   // time
                                   // of the last sound event
int _gapBetween2Sounds = 1000;     // gap between 2 sound
                                   // events

//**********************************************************/
// Step-2: INITIALIZE I/O PARAMETERS
//**********************************************************/
void setup ()
{
  // configure the pins to be used for I/O
  pinMode (_pinD10, INPUT) ; // input from the sound sensor
  pinMode(_pinD7, OUTPUT);   // output to relay
}

//**********************************************************/
// Step-3: MAIN PROGRAM
//**********************************************************/
void loop ()
{
  _soundLevel = digitalRead (_pinD10) ; // read sound level
  if (_soundLevel == LOW)               // if a sound occurs
  {
    if (_bRelayState == false)          // if relay is in
    {                                   // OPEN state
      //if event is after at least 1 second of the previous
      // event
      if( (millis()-_lastSoundEventTime)> _gapBetween2Sounds)
      {
        _lastSoundEventTime = millis(); // record event time
        _bRelayState = true;
        //Switch ON the Relay
        digitalWrite(_pinD7, 0);
      }
```

```
      }
    if (_bRelayState == true) //if relay is already ON
    {
      //if event is after at least 1 second of the previous
      // event
      if( (millis()-_lastSoundEventTime)> _gapBetween2Sounds)
      {
        _lastSoundEventTime = millis(); // record event time
        _bRelayState = false;
        //Switch OFF the Relay
        digitalWrite(_pinD7, 1);
      }
    }
  }
}
```

The first two steps of the sketch are self-explanatory in that they are used to setup initialization parameters and variables and specify the input/output modes of the pins used by the embedded code. The main program keeps processing any sound events that occur. Upon detecting the first sound event, the light bulb is switched ON.

For the sound detector, it is very important to know how to handle incoming signals while working with a device that sends a continuous stream of signals. Otherwise too many sound events will be captured and processed by the program and the light bulb will start getting switched ON and OFF too quickly in succession. To avoid this situation, the following statement has been used to track the last time the sound started:

```
    if ( (millis() - _lastSoundEventTime) > _gapBetween2Sounds)
```

The preceding code is written in such a way that there will be a mandatory gap of 1 second before the program accepts a sound event for processing. So, in essence, if someone claps their hands twice within a 1 second time window, then the second clap will be ignored. There should be a gap of at least one second between two subsequent sound events. Try to change the value of the variable gapBetween2Sounds to a higher value and see how the sketch behaves.

Part 2 - Actual prototype for sound activated light bulb controller

In this section, we will use a real AC powered device. All that needs to be done is to replace the LED and the DC power supply with an electric bulb and AC mains supply. Do not attempt this if you are a novice with electricity and do not know how to handle AC mains.

 Caution: Remember that you will be working with AC mains. Improper use of AC mains can be fatal.

- All AC power supply should be switched OFF while building the circuit with the relay
- DO NOT keep any part of the circuit on a conductive surface
- DO NOT touch the circuit when operational
- Wear insulated protective gear while working with this prototype
- Make sure there are no short circuits

Although the setup and sketch has been tested carefully, the author and/or the publisher shall not be held liable for any damage caused directly or indirectly resulting from the use of this discourse.

For building this prototype, the following parts will be required:

- Arduino Uno R3
- USB connector
- One 4 battery holder + four 1.5V AA batteries
- One sound detection sensor module
- One 5V 10A AC power relay
- One AC powered light bulb (with holder)
- Copper wires as required

Purchase the relay based on the AC mains rating in your country. For example, in some countries the voltage rating is 220V, whereas in other countries it may be 110V.

Follow the schematic diagram shown below to build the circuit for interfacing the electric bulb with the Arduino. Remember to disconnect the AC mains while assembling the circuit.

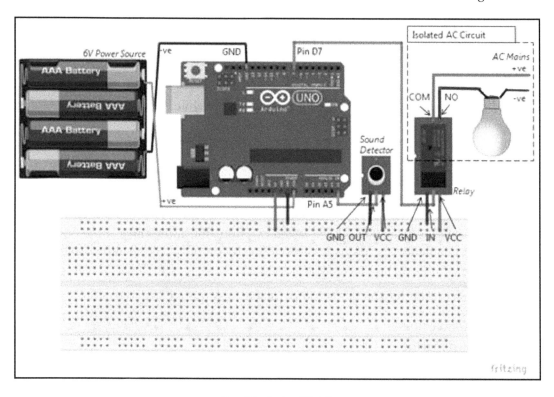

Figure 5: Sound activated light bulb

As shown in the preceding figure, the sound detector to Arduino Uno connections are tabulated as follows:

Arduino Uno pin	Sound sensor pin
Analog I/O Pin A5	OUT
5V	VCC
GND	GND

Table 5: Sound sensor to Arduino connections

As shown in the preceding figure, the relay to Arduino Uno connections are tabulated as follows:

Arduino Uno pin	Relay Port/Jack/Pin
Digital I/O Pin 7	IN
5V	VCC
GND	GND

Table 6: Relay to Arduino connections

The relay to AC power and lamp/bulb connections are tabulated as follows:

Relay device pin	AC Power/Device
COM	To +ve power supply terminal
NO	To Lamp/Bulb

Table 7: DC Devices to relay connections

The positive terminal of the power supply should be connected to the COM terminal of the relay device. The negative terminal of the power supply should be connected to the other end of the lamp.

Future inspiration - Automatic room lights

Based on what we have learnt in this chapter and in Chapter 6, *Day 4 - Building a Standalone Device* previously, the following figure is a simple idea to automate the lights at your home:

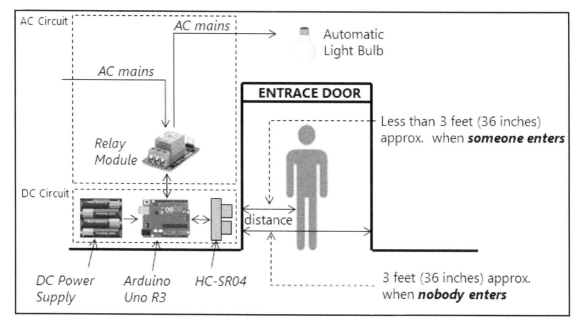

Figure 6: Automatic room lights

As depicted in the preceding figure, the Arduino is interfaced with your room light via a Relay module in order to keep the DC circuit separate from the high voltage AC mains. Utilize what we learnt in this chapter for achieving the Relay module interfacing.

The DC part of the circuit has an HC-SR04 ultrasonic sensor and has been powered from an external battery. Use the knowledge you gathered in Chapter 6, *Day 4 - Building a Standalone Device* for the basic wiring and sketch for using an ultrasonic sensor with the Arduino.

Place the HC-SR04 ultrasonic sensor strategically on one side of the entrance door to a room. Let us say the door is 3 feet (36 inches) in width. The ultrasonic sensor should be placed in such a way, that normally the ultrasonic sensor should only read the width of the door. If nobody enters through the door then the ultrasonic sensor will always return a distance of 3 feet (36 inches) approximately (do not worry about precision at this point). However, as soon as someone enters through the door, the ultrasonic sensor will immediately read a reduced distance (which will be less than 36 inches). You can reuse the sketch from `Chapter 6`, *Day 4 - Building a Standalone Device* and introduce an `if` statement to capture the event when someone enters through the door, as shown in the following code:

```
// Convert the distance to centimeters
distance = (duration/2) / 29.1;

// you may need to tweak the number 30 below
// depending upon the width of your door
// and also depending upon what object will be entering the room
if (distance < 30)
{
  // trigger the AC-DC Relay device to ON state
  // so that the interfaced room light is switched ON
}
```

The factor 29.1 shown in the preceding code has been calculated based on the speed of sound in air, explained as follows:

$$343 \ m/sec = 34300 \ cm/sec = 0.0343 \ cm/micro\text{-}sec = 1/29.1$$

So as soon as someone tries to enter through the door, they will cut into the line of measurement of the ultrasonic sensor. This will result in the ultrasonic sensor returning a lower value than the width of the door. That is when the sketch will detect that someone is trying to enter the room and will trigger the Relay to ON position, thus resulting in switching ON the room light.

Try the following

Let us try the following things before proceeding to the next chapter:

- Increase the value of the `_gapBetween2Sounds` variable to a larger value and see how your prototype behaves

- Decrease the value of the _gapBetween2Sounds variable to a larger value and see how your prototype behaves
- Set the value of the _gapBetween2Sounds variable to zero (0) and see how your prototype behaves

Things to remember

Remember the following important points about what we learnt in this chapter:

- A relay is used to interface the Arduino board (or other micro-controller devices running on DC power) with AC powered equipment
- The positive end of the AC electric supply is connected to the COM port of a relay
- The positive end of the AC powered equipment is connected to the NO port of a relay
- The IN/IN1 port of a relay is used to accept a signal from Arduino, based on the signal the relay either switches ON (closed state) or OFF (open state)
- The `digitalRead` function is used to read input signals via Arduino's digital pins
- The `digitalWrite` function is used to send output signals via Arduino's digital pins.

Summary

In this chapter, we learnt how to build a prototype for interfacing AC powered devices with the Arduino board as the controlling device. In the first part of the chapter, we learnt how to use a Relay device with the Arduino. In the second part we used the Relay device with an electric bulb. Eventually we used a sound detector to operate the Relay (indirectly the bulb) based on ambient sound.

In the next chapter we will learn how to use Infrared signals for wireless communications with the Arduino. We will look at transmitting as well as received data using Infrared transmitter and receiver components.

9

Day 7 - The World of Transmitters, Receivers, and Transceivers

"If you can see the invisible, you can achieve the impossible."
- Shiv Khera

Exciting as it sounds, today we will enter the world of transmitting and receiving data over the air. As you read through, this chapter will unravel and demystify some embedded world techniques used for transmitting and receiving data from one device to another by using Infrared wireless signals. In the next chapter, we will learn to use radio frequency. Using this information, we will be able to move to the next level of Arduino prototyping by going wireless!

Things you will learn in this chapter:

- Understanding Infrared communications
- Hacking into an existing remote control
- Building an Infrared receiver device
- Using IR receiver TSOP1738/TSOP1838
- Using IR receiver SM0038
- Building an Infrared transmitter device
- Using IR transmitter LED
- Controlling Arduino projects

Understanding Infrared communications

Infrared (**IR**) communications are very commonly used in the remote-control industry for transmitting and receiving wireless signals from one device to another device. To understand better, think of the ambient light around us; it contains various types of light, where each type of light has a different wavelength.

The study of light itself is a large topic; however, it would be good to know a few basics when trying to work with Infrared communications. Light that is visible to the human eye falls in the wavelength range of 400 nm to 700 nm (where nm is the abbreviation for nanometer) - this is known as the visible spectrum of ambient light. IR light is not visible to the human eye and falls in the wavelength range of 700 nm to 1 mm. Infrared light has a longer wavelength and is not visible to the human eye. Since IR light is not visible to the human eye, it does not interfere with our senses even when IR communications are going on right in front of us.

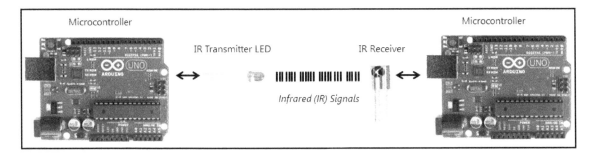

Figure 1: Typical infrared communications setup

Infrared communication devices usually consist of the following parts (as depicted in the preceding figure):

- A transmitting device (transmitter LED + microcontroller)
- A receiving device (receiver unit + microcontroller)

Usually, on the transmitting side, a microcontroller sends electronic pulses to an IR transmitter LED. The transmitter LED in turn emits, via IR driver software, IR wavelength conformant signals. While transmitting the IR signals, the point to note is the frequency at which the signals are transmitted.

Infrared communication frequency

By frequency of transmission, we mean how many times the signal goes to a logical HIGH and LOW within just 1 second.

 The most common frequency to transmit IR signals is 38 KHz (Kilo Hertz), in other words, the IR LED will be triggered 38,000 times in 1 second by the microcontroller. In response to the microcontroller pulses, the IR LED will also emit quick and short IR waves 38,000 times in 1 second. These 38,000 IR waves will be carried through the air and received by the IR receiver hardware.

Apart from the 38 KHz frequency there are other frequencies also in use in the industry. We will learn using IR communications with Arduino using this most commonly used and implemented 38 KHz frequency.

On the receiver side, usually there is an IR receiver module, with in-built hardware, that receives and de-modulates the incoming IR signals. The de-modulated IR signals are made available on an output pin of the IR receiver. This output pin is attached to a microcontroller that can read, via IR driver software, the IR signals.

In the case of the Arduino, the IR driver software for sending and receiving IR signals is available in the form of an Arduino library that we will learn to use in the following sections of this chapter.

Infrared communication protocol

While working with Infrared signals, we should also know that there are several Infrared communication protocols available on the market. These protocols are collectively referred to as **Consumer Infrared (CIR)**.

Unlike internet TCP/IP packet transmissions, Infrared transmissions are an unregulated area. Over the years, electronics device manufacturers have defined and implemented their own protocols to transmit and receive Infrared signals. This was done to avoid interference in the cross operation of electronic devices placed side by side. For example, a Phillips DVD player with a Samsung TV - operating their individual remote controls should not interfere with the other device accidentally. In other words, we will not be able to increase the volume of a Samsung TV set with a remote control that works with a Panasonic TV set and vice-versa. Thus, IR protocols play a very important role.

For example, many Japanese manufacturers follow the NEC protocol for Infrared communications, whereas, European players like Phillips would have their own standard Infrared protocol. Some manufacturers like Sony have developed a proprietary protocol called **S-Link**. Similarly, there are many protocols in use today. IR protocols are an advanced topic, hence we will focus on the basics in this chapter.

 You must use a compatible protocol for transmitting IR signals. For example, if trying to transmit IR codes to a Panasonic device, then use the Panasonic protocol. If using an LG device, then use the LG protocol, and so on.

In the latter half of this chapter, we will learn using appropriate protocols for transmitting IR signals to control devices of our choice.

Hacking into an existing remote control

As we proceed through this chapter, we will take the example of a common TV remote control and learn how to hack into it and use it with an Arduino board. The general process of hacking into an existing remote control set has been explained in the following main sections:

- Using IR receiver TSOP Series IR receivers
- Building an Infrared transmitter device

We will use the knowledge in the following sections to receive the TV remote control's IR codes by using the IR receiver sketch. Later we will re-use the IR codes by transmitting them using the IR transmitter sketch and independently control the TV directly from the Arduino (without using the TV remote control). The same techniques can be used to read the IR codes of other remote control sets and also to transmit IR codes for controlling your own Arduino project remotely.

Building an Infrared receiver device

Building an IR receiver device is intuitively easier and will be our first step towards learning IR communications using Arduino. For building a basic Infrared receiver device we will use the following parts:

- One Arduino Uno R3
- One USB cable
- One IR receiver (TSOP family of IR receivers)
- One pc. 1K Ohms resistor
- Some male-to-male jumper wires

In the following sections, we will see examples of two popular IR receivers from the TSOP family of IR receivers and the SM0038 IR receiver. The IR receivers have in-built circuitry for receiving and decoding the IR signals. No additional circuitry is required for decoding the received IR signals. All you need to do is plug in the proper breadboard connection with the Arduino.

The first example will be a detailed example. We will use the TSOP1738/TSOP1838 IR receiver to build an Arduino-based device to receive IR signals from any remote control. You can use this device to detect and read incoming IR signals from any remote control that operates at a frequency of 38 KHz. Most remote controls that are commonly used to control general electronic gadgets like car DVD players, room light, robotic vehicles and so on use 38 KHz and should work with this example.

The second example will be a short guidance on using the SM0038 IR receiver to receive 38 KHz IR signals. You can use this receiver with almost any remote control lying around the house. In this case, we will reuse the sketch from the first example. The only change will be in the pin out of the SM0038 IR receiver.

Both these examples are quite similar to one another, the only difference being the IR receiver pin outs. These examples have been chosen to demonstrate the versatility of using the Arduino platform with different families of IR receivers.

The Arduino Infrared library

Before beginning with the examples, we must quickly setup the Arduino programming environment for working with Infrared signals. This part is a bit tricky, because we are going to manually install an Arduino library for working with Infrared communications.

The ArduinoIDE (version 1.6.9 has been used in this book) comes preloaded with an Infrared remote-control library. However, this library is designed for Arduino compatible robotic vehicles that are manufactured separately - it does not work directly with normal IR receivers. For straightforward working with IR devices, there is a very popular open source IR library for Arduino that was written by Ken Shirriff. We will use the Ken Shirriff IR library instead of the IR library that shipped with the Arduino IDE.

 If you try to compile the IR sketches mentioned in this book directly in the Arduino IDE, without first installing the IR library by Ken Shirriff, then the following compilation error will be seen:

```
C:\Program Files
(x86)\Arduino\libraries\RobotIRremote\src\IRremoteTools.cpp:5:16: error:
'TKD2' was not declared in this scope
int RECV_PIN = TKD2; // the pin the IR receiver is connected to
```

In order to avoid the above problem, follow the steps mentioned below to remove the IR library that shipped with the Arduino IDE and replace it with Ken Shirriff's IR library.

First let us understand how to remove the existing library that shipped with the Arduino IDE. While removing the existing library, retain the removed folder in another location (outside the `Program Files` location mentioned as follows) so that if needed you may be able to use it in future. Alternately you may download it anytime from `https://github.com/arduino-libraries/RobotIRremote`.

1. Browse to the Windows path `C:\Program Files (x86)\Arduino\libraries` (as shown in the following screenshot):

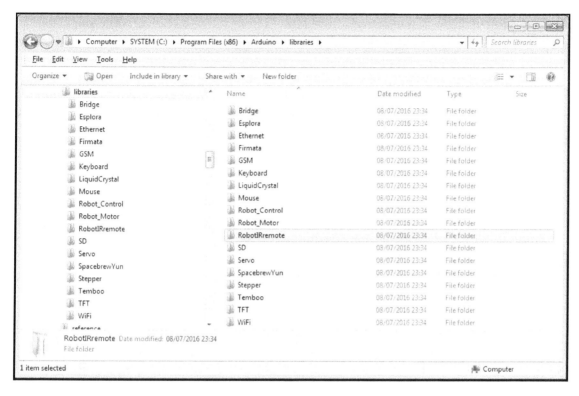

Figure 2: Remove preloaded IR library shipped with Arduino IDE

2. Locate the folder titled `RobotIRRemote` (as shown in the preceding picture).

3. Next, cut the `RobotIRRemote` folder from this location.

4. Then, paste the `RobotIRRemote` folder in any other location on your computer. The location should be outside the above mentioned folder.

Performing the preceding steps will ensure that the existing library gets removed from the Arduino IDE. Now if we try to compile any of the sketches that are included in this chapter, we will receive the following errors:

```
C:\Users\a603209\Desktop\trial\IR_Remote\IR_Remote.ino:1:22: fatal error:
IRremote.h: No such file or directory
#include <IRremote.h>
compilation terminated.
```

The preceding errors will confirm that we have successfully removed the existing library from the Arduino IDE system.

The last step will be to download and install the IR library by Ken Shirriff. The steps are outlined as follows for reference:

1. Browse to the following GitHub location `https://github.com/z3t0/Arduino-IR remote/releases`:

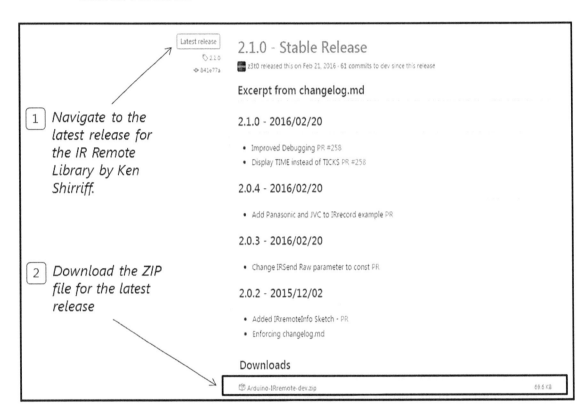

Figure 3: Ken Shirriff IR library download from GitHub

2. Download the latest stable release, as shown in the preceding picture. 2.1.0 was the latest release at the time of writing this book. You should download the latest library at the time of reading this book.

3. After downloading the ZIP file, unzip the contents in a folder with the title, as shown in the following figure. Keep the name of the folder the same as the name of the ZIP file for easy reference for the future:

Figure 4: Downloaded and unzipped folder Arduino-IR Remote-master

4. After extracting the folder, simply copy the entire folder (as shown in the following figure):

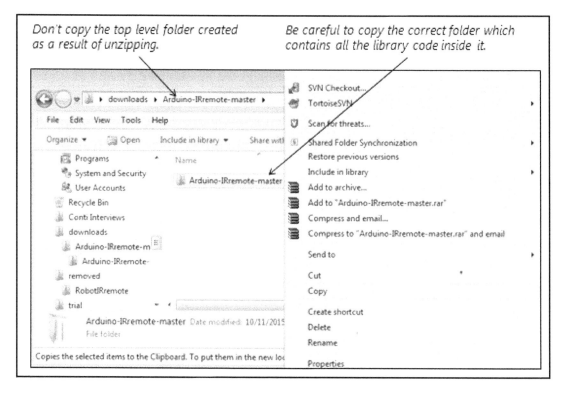

Figure 5: Copy newly downloaded library folder

5. Then paste it inside the Arduino libraries path `C:\Program Files (x86)\Arduino\libraries` (as shown in the following figure):

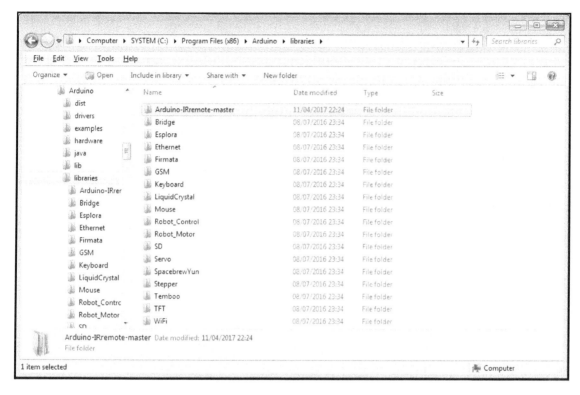

Figure 6: Ken Shirrif's IR library installed

That's it! We have now successfully installed the IR library that we will use with our examples in this chapter. Now if you try to compile any of the sketches provided in this book, you will notice that the compilation will succeed, without any errors.

Using IR receiver TSOP series IR receivers

The TSOP family of IR receivers comes in different varieties, mainly based on the IR frequency that they can detect. Two of the most commonly used IR receivers are TSOP1738 and TSOP1838; both are sensitive to 38 KHz. They have all of the required internal circuitry to receive IR signals of 38 KHz. In this section, we will use an ordinary universal remote control and receive its transmitted signals using the TSOP1738 and TSOP1838.

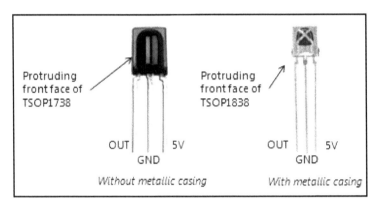

Figure 7: TSOP1738/TSOP1838 pinout

The TSOP IR receivers shown in the preceding figure typically have a protruding front face. This protruding face of the IR receiver should be free of any obstruction so that it can receive the IR signals. Typically, the TSOP1738 comes without a metal casing, whereas the TSOP1838 models usually ship with a metallic casing.

The TSOP IR receivers shown in the preceding figure have three legs (from left to right when the protruding face is towards us):

- **Out**: This is the output pin of the IR receiver. This pin is connected to a microcontroller unit. The IR receiver sends the decoded IR signal through this pin. In our case, this pin will be connected to the Arduino board. Our Arduino sketch will read the decoded IR signal from this pin.
- **GND**: This pin will be connected to the ground pin of the Arduino board.
- **5V**: this pin will be connected to the 5 volt power supply pin of the Arduino.

The parts required for building the IR receiver prototype are listed as follows:

- One consumer remote control set (a normal TV remote can be used)
- One Arduino Uno R3
- One USB A to USB B cable

- One TSOP1738 IR receiver (you can substitute with TSOP1838 also)
- One piece 1K Ohms resistor
- Some male-to-male jumper wires

Once all the mentioned preceding parts have been assembled, go ahead and build the circuit shown in the following breadboard diagram:

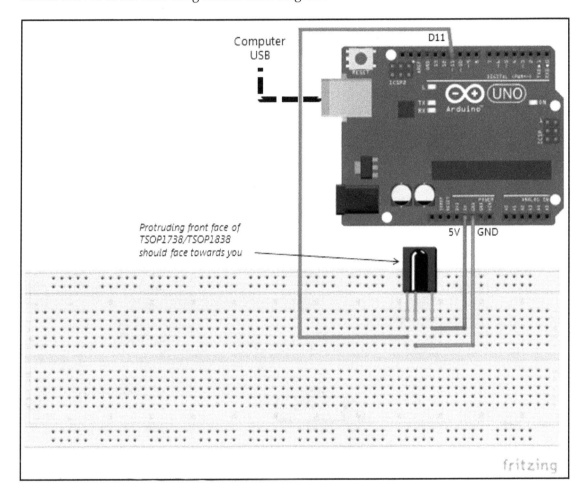

Figure 8: IR receiver using TSOP1738 or TSOP1838

In the preceding diagram, the IR receiver's protruding face should face towards you in the circuit setup. TSOP1738 has been used in the preceding setup. However, we can simply substitute it with TSOP1838, without any change in connections.

The accompanying sketch for the IR receiver setup is provided below for reference. It is based on the sketch using the Arduino Infrared library written by Ken Shirriff. A similar sketch is available in the public domain:

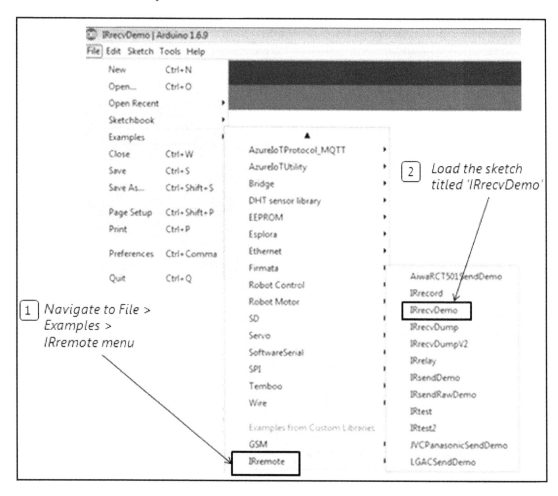

Figure 9: Load IR receiver sketch shipped with Ken Shirriff library

The sketch used in this example can be loaded by simply navigating to the Arduino IDE menu `File|Examples|IRRemote|IRRecvDemo` (as depicted in the preceding figure). For our understanding, the above sketch has been slightly modified with some additional comments. All credits go to Ken Shirriff for the original code. You can download the sketch from the online location mentioned in `Chapter 1`, *Boot Camp*, of this book also:

```
// This sketch has been modified based on the original sketch
// written by Ken Shirriff
//*******************************************************/
// Step-1: CONFIGURE VARIABLES
//*******************************************************/
#include <IRremote.h>
int RECV_PIN = 11;            // Specify the pin to read the
                              // input from the IR Receiver OUT pin
IRrecv irrecv(RECV_PIN);      // Define the object for the
// IR receiver
decode_results results;       // Define an object to store results


//*******************************************************/
// Step-2: INITIALIZE I/O PARAMETERS
//*******************************************************/
void setup()
{
  Serial.begin(9600);         // Start serial communications
  irrecv.enableIRIn();              // Start the receiver
  Serial.println("Receiver Started...");
}


//*******************************************************/
// Step-3: MAIN PROGRAM
//*******************************************************/
void loop()
{
  if (irrecv.decode(&results))
  {
    Serial.println(results.value, HEX);
    irrecv.resume(); // Receive the next value
  }
}
```

Now let us quickly understand the above sketch. To start with, we are using the IR library written by Ken Shirriff:

```
#include <IRremote.h>
```

All the necessary functions to manage IR communications are already coded in the above library. The next step is to specify the Arduino pin that will read the decoded IR signals from the IR receiver's OUT pin and configure the variables for the IR receiver object:

```
int RECV_PIN = 11;        // Specify the pin to read the
                          // input from the IR Receiver OUT pin
IRrecv irrecv(RECV_PIN);  // Define the object for the
                          // IR receiver
```

In the `setup()` function, you will notice that the IR receiver software has been started by calling the appropriate function:

```
irrecv.enableIRIn();      // Start the receiver
```

After the receiver object has been setup, the `loop()` function keeps sampling for incoming IR signals. As soon as an IR signal is received, the value of the decoded signal is stored in the `results` variable:

```
if (irrecv.decode(&results))
```

Thereafter, the value is printed to the Serial Monitor window. In the following line of code, we are printing the hexadecimal value corresponding to the decoded IR signal. It is worthwhile to note that we can also print the values of the decoded IR signals in other formats such as decimal, if needed.

```
Serial.println(results.value, HEX);
```

After printing the value, the program immediately resumes listening to the next incoming IR signal from the IR receiver's OUT pin using the following statement:

```
irrecv.resume(); // Receive the next value
```

Thus, the sketch keeps running. Load the sketch onto your Arduino board and launch the Serial Monitor window:

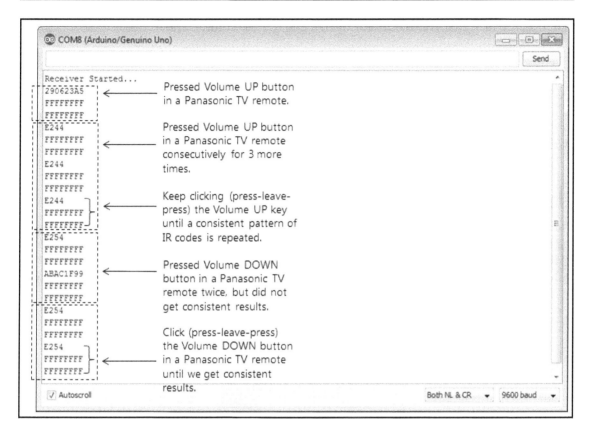

Figure 10: Reading the IR codes from a TV remote

Point a remote control at the IR receiver and press any button. As soon as any button is pressed, the corresponding IR signal code will get displayed on the screen. The above Serial Monitor window displays IR signal codes from a Panasonic TV remote. You can try this with multiple brands of remote controls.

Based on the above results captured in the Serial Monitor window, it seems the IR signal HEX codes for increasing the TV volume are:

- E244
- FFFFFFFF
- FFFFFFFF

Whereas the IR signal HEX codes for decreasing the TV volume are:

- **E254**
- **FFFFFFFF**
- **FFFFFFFF**

Later in this chapter, we will learn how to send these codes and control the TV unit using an Arduino board and an IR transmitter LED.

Using IR receiver SM0038

There is another popular IR receiver that can be used with the Arduino sketch that we just learned in the previous example. This IR receiver is known as the SM0038. We can use the SM0038 as-is with the sketch in the previous section on using a TSOP1738/TSOP1838. The only difference will be the wiring, since the SM0038 legs are arranged differently, as shown in the following figure:

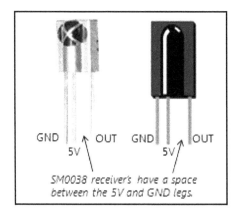

Figure 11: SM0038 IR receiver pinout

The SM0038 IR receivers shown in the preceding figure have three legs (from left to right when the protruding face is towards us):

- **GND**: This pin will be connected to the ground pin of the Arduino board.
- **5V**: This pin will be connected to the 5 volt power supply pin of the Arduino.
- **Out**: This is the output pin of the IR receiver. This pin is connected to a microcontroller unit. The IR receiver sends the decoded IR signal through this pin. In our case, this pin will be connected to the Arduino board. Our Arduino sketch will read the decoded IR signal from this pin.

For using the SM0038 with the Arduino, simply change the wiring in the breadboard diagram shown for TSOP1738/TSOP1838. Also, we can reuse the same sketch that we used with TSOP.

Building an Infrared transmitter device

Now that we have learnt how to build an IR receiver device, let us understand the basics of transmitting IR signals. For transmitting IR signals, we will have to use special IR Transmitter LEDs. IR transmitter LEDs come in various colors: blue, transparent, and so on. For the example in this section, we will use a blue IR LED.

Using an IR transmitter LED

The IR transmitter LEDs look similar to normal LEDs, but are slightly larger than the normal LEDs:

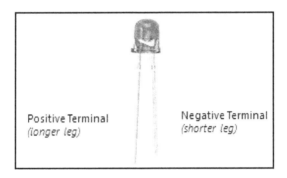

Figure 12: A typical IR transmitter LED

An IR transmitter LED has two legs (terminals) as shown in the preceding figure:

- Positive terminal (longer leg), this pin will be connected to the 5-volt power supply pin of the Arduino.
- Negative terminal (shorter leg), this pin will be connected to the GND pin of the Arduino, via a transistor.

The parts required for building the IR transmitter prototype are listed as follows:

- One Arduino Uno R3
- One USB A to USB B cable
- One IR transmitter LED (Blue/Transparent/other color)
- One NPN transistor (such as a N2222/BC547 general purpose NPN transistor)
- 470 Ohms resistance (suggested: 220 + 150 + 100 = 470 Ohms)
- Some male-to-male jumper wires

Once all the mentioned preceding parts have been assembled, go ahead and build the circuit shown in the following breadboard diagram:

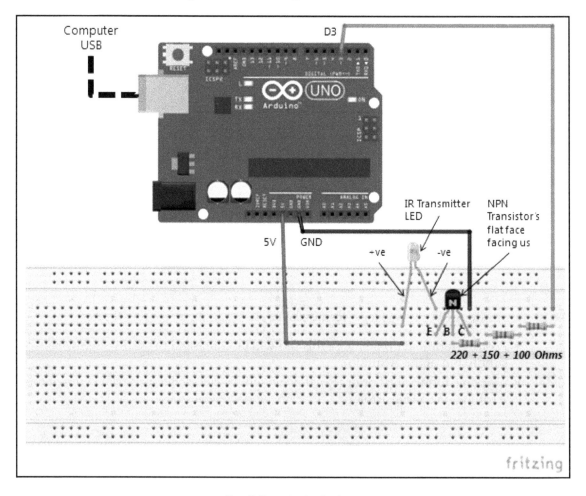

Figure 13: IR transmitter breadboard setup

The accompanying sketch for the IR transmitter setup is provided below for reference. This sketch can be downloaded from the online location for this chapter mentioned in `Chapter 1`, *Boot Camp*. This sketch has been designed with a Panasonic TV remote. It increases the TV volume every time the letter 'l' or 'L' is input through the Serial Monitor window. The TV volume will be decreased every time the letter 'a' or 'A' is entered in the Serial Monitor window:

```
//***********************************************************/
// Step-1: CONFIGURE VARIABLES
//***********************************************************/
#include <IRremote.h>
IRsendMy_Sender;


//***********************************************************/
// Step-2: INITIALIZE I/O PARAMETERS
//***********************************************************/
void setup()
{
  Serial.begin(9600);
}


//***********************************************************/
// Step-3: MAIN PROGRAM
//***********************************************************/
void loop()
{
  char cmd = Serial.read();

  // Enter letter 'l' or 'L' to increase sound
  if ( cmd == 'l' || cmd == 'L' )
  {
    //My_Sender.sendNEC(0xE244, 16); // Slow response in TV
    My_Sender.sendNEC(0xE244E244, 32);   // Fast response in TV
    My_Sender.sendNEC(0xFFFFFFFF, 32);
    My_Sender.sendNEC(0xFFFFFFFF, 32);
  }

  // Enter letter 'a' or 'A' to decrease sound
  if (cmd == 'a' || cmd == 'A')
  {
    //My_Sender.sendNEC(0xE254, 16);    // Slow response in TV
    My_Sender.sendNEC(0xE254E254, 32); // Fast response in TV
    My_Sender.sendNEC(0xFFFFFFFF, 32);
    My_Sender.sendNEC(0xFFFFFFFF, 32);
  }
}
```

Now let us quickly understand the preceding sketch. The initialization and configuration is very similar to the sketch we learnt for receiving IR signals. The `loop()` function contains the main logic for increasing and decreasing the volume of the TV set. First, it scans the input from the Serial Monitor window:

```
char cmd = Serial.read();
```

If the letter 'l' or 'L' has been entered, then the sketch executes the code for transmitting the IR codes for increasing the TV set's volume. You will notice that one line of code is commented out. This commented line of code sends only four HEX digits (E244) to the TV, whereas the uncommented code sends eight HEX digits (244E244). If we try to send the four HEX digits then by trial and error, we will notice that the TV set responds slowly as compared to sending the eight HEX digits. This is due to the way the Panasonic TV set accepts the IR codes based on its intrinsic protocol and the behavior may differ across models and brands. So, if you are using another brand of TV set then you may have to perform some trial and error on your own:

```
if ( cmd == 'l' || cmd == 'L' )
{
  //My_Sender.sendNEC(0xE244, 16); // Slow response in TV
  My_Sender.sendNEC(0xE244E244, 32);   // Fast response in TV
  My_Sender.sendNEC(0xFFFFFFFF, 32);
  My_Sender.sendNEC(0xFFFFFFFF, 32);
}
```

Similarly, if the letter 'l' or 'L' has been entered, then the sketch executes the code for transmitting the IR codes for increasing the TV set's volume:

```
if (cmd == 'a' || cmd == 'A')
{
  //My_Sender.sendNEC(0xE254, 16);    // Slow response in TV
  My_Sender.sendNEC(0xE254E254, 32); // Fast response in TV
  My_Sender.sendNEC(0xFFFFFFFF, 32);
  My_Sender.sendNEC(0xFFFFFFFF, 32);
}
```

 The important point to note is the format of the IR send function. For example, when using the `sendNEC(code, length)` function, for sending a HEX code, always prefix it by `0x`. For each HEX digit, the `length` parameter should be increased by 4.

So, when sending four HEX codes, the usage would be:

```
My_Sender.sendNEC(0xE254, 16);
```

Whereas when sending eight HEX codes, the usage would be:

```
My_Sender.sendNEC(0xE254E254, 32);
```

Controlling Arduino projects

In the previous section, we learnt how to transmit IR signals to common household devices such as a TV set. The limitation in trying to control household devices is the fact that even if we can decode the IR codes from their remote-control sets, these devices might refuse to accept an IR code sent by our sketches. This is because these devices have their proprietary protocols. If the IR codes sent do not match with the specific device protocols then it will not work. In the preceding example, we were lucky because most Japanese manufacturers use the NEC protocol. Therefore, we were able to transmit the IR codes and control the TV set's volume.

The above limitation does not apply when we try to control our Arduino-based devices. For controlling Arduino Projects with a remote-control, we can use any remote-control lying around our household or office or we can also procure a brand new universal remote-control from the online market place.

There are two important limitations to consider when trying to control your Arduino projects with Infrared. The first limitation is that Infrared signal propagation works only if there are no physical obstructions between the IR transmitter and IR receiver. In other words, we need an unobstructed line-of-sight for Infrared signals to propagate through the air. This is due to the nature of Infrared red light. IR signals will not bend around corners or the obstacles to reach the target.

The second important thing to keep in mind is the short range of around 25 to 30 feet. Depending upon some varying factors such as power source quality and transmitter build quality, the range of the IR signals will vary. So if you want to control something within 25 to 30 feet, then Infrared might be the right choice.

All that matters is to receive an IR code and then based on that IR code invoke an appropriate function in our device. As a simple example, we could reuse the Panasonic TV remote's volume control IR codes to increase or decrease the speed of a DC motor. An indicative Arduino sketch for doing so is shown below:

```
void loop()
{
  if (irrecv.decode(&results))
  {
    // IR code for Volume UP button on Panasonic remote
    // Replace the value 0xE244 below with the IR code in the
```

```
          // remote control being used.
          if (results.value == 0xE244);
          {
            // invoke function to increase DC motor speed
            increaseSpeed();
          }

          // IR code for Volume DOWN button on Panasonic remote
          // Replace the value 0xE254 below with the IR code in the
          // remote control being used.
          if (results.value == 0xE254);
          {
            // invoke function to decrease DC motor speed
            decreaseSpeed();
          }

          irrecv.resume(); // Receive the next value
        }
      }
```

As we can see in the above code snippet, all we have to do is detect and get the decoded IR signal incoming from a remote-control. After the signal has been decoded and read, the decoded IR code is stored in the `results` variable. Based on the `results.value` the appropriate function should be invoked.

The above example is based on the IR receiver device sketch that we learnt in the first part of this chapter. You can apply the same format of the preceding sketch snippet to control any Arduino project wirelessly!

Transceivers

So far, we have seen the TSOP and SM devices as examples of transmitters. We also saw IR LEDs as transmitters. There is one more category of components known as transceivers. As the name suggests, transceivers are a combination of a transmitter and a receiver built-in together as a single electronic component or module.

We have already seen an example of a transceiver in Chapter 6, *Day 4 - Building a Standalone Device*. The HC-SR04 Ultrasonic sensor is actually a transceiver. It transmits ultrasonic waves and also receives the reflected ultrasonic waves.

Similarly, just like the HC-SR04 Ultrasonic sensor we can pair an IR Transmitter LED with a photodiode to build a transceiver component to measure distances. This type of Infrared transceiver is used as a proximity sensor of various robotic vehicles to sense obstacles and navigate itself.

Try the following

At the end of this chapter, we arrive at the point where we must apply what we have learnt so far. So, let us dive into the following exercises.

Utilize the information in `Chapter 8`, *Day 6 - Using AC Powered Components* for the relay operated electric light bulb and then try to fit in an IR receiver in that circuit. After you have done that, make the necessary changes in the sketch to operate the light using a remote control.

Things to remember

Remember these important points while using the Arduino platform in your future projects.

The most common frequency to transmit IR signals is 38 KHz (Kilo Hertz).

The TSOP1738, TSOP1838 and SM0038 are responsive to, and can detect and decode, IR signals of 38 KHz.

If you want to decode IR signals of a higher frequency than you must use a compatible IR receiver that is responsive to that frequency.

Use the Arduino IR Library written by **Ken Shirriff**, to work with generic IR components. You must remove the pre-existing IR Remote Library before installing the IR Library by Ken Shirriff.

When using Ken Shirriff's Arduino Library, on an Arduino Uno, only digital I/O Pin 3 can be used for transmitting IR signals. This is because the library internally changes the frequency of pin 3 and utilizes it to send the signals.

The distance up to which an IR LED can transmit will be limited by the power supply. Use an external battery power supply for better range, as Arduino's pins cannot supply the amount of power that an external battery can provide.

Summary

In this chapter, we concentrated on understanding the basics of Infrared communications with the Arduino platform. We looked at some commonly used IR receiver components such as TSOP1738, TSOP1838, and SM0038 that can be used for detecting and decoding incidental IR signals. During this lesson, we saw how to capture the IR code from a remote control.

In the second part of the chapter, we learnt how to work with IR Transmitter LEDs for transmitting IR codes. We also learnt how to send IR codes using the Arduino platform to a TV set and control its volume.

In the last part of the chapter, we looked at a generic way of using remote controls to control our Arduino projects. Using the basic fundamentals that you learnt in this chapter, you should attempt to build some more examples on your own.

In the next chapter, we are going to learn how to build Radio Frequency enabled devices. Learning to build Radio Frequency enabled devices will be our next step to start building more wireless devices with the Arduino platform.

10

Day 8 - Short Range Wireless Communications

"Radio is powerful not because of the microphones, but the one who sits behind the microphones."
- Ernest Agyemang Yeboah

In this chapter, we will learn different techniques for performing short range wireless communications using the Arduino platform with **radio frequency** (**RF**) technology. It is generally perceived that radio waves can travel very long distances, which is true. However, there are different types of radio waves. Some radio waves travel across the globe and beyond this planet, while some can travel only a few hundred meters.

As we read through this interesting chapter, we will discover how to use some popular RF transceiver chips for sending and receiving radio waves.

You will learn the following topics in this chapter:

- Building an RF device
- Using the nRF24L01 2.4 GHz ISM band RF transceiver chip with Arduino
- Transmitting RF waves
- Receiving RF waves
- **Machine to Machine** (**M2M**) communications using RF chips
- Using the HC-05 Bluetooth chip with Arduino

Building a radio frequency device

Radio frequency waves are part of the electromagnetic spectrum. Usage of radio frequency waves is strictly regulated by laws. We have all heard about telecom operators bidding for various spectrums. They are actually entering into an agreement with the governments around the world that they be allocated a particular band in the electromagnetic spectrum surrounding the earth. The governments in turn decide who will be allowed to use which band in the electromagnetic spectrum. Post allocation, a telecom operator can exclusively use the allocated band for their telecommunication operations. The same applies to radio station channels airing on particular radio frequencies.

There are various types of radio frequency waves depending upon their frequencies between the ranges of 3 KHz to 300 GHz. There is a special radio frequency that has been set aside worldwide, for use by the **Industrial, Scientific and Medical (ISM)** community. This is officially known as the 2.4 GHz ISM band. In this chapter, we will learn how to use 2.4 GHz RF chips for transmitting and receiving short range radio waves. Practically, short range would mean not more than 30-50 meters.

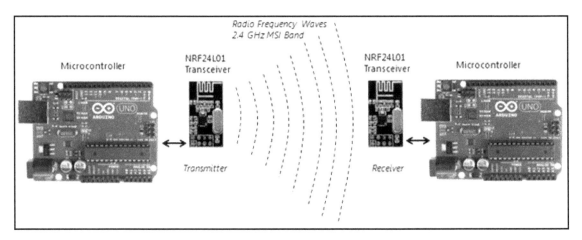

Figure 1: Overview of a typical RF transmitter-receiver pair

Typically, RF-based solutions use an RF capable chip or module together with a microcontroller. As shown in the preceding figure, at one end we have the transmitting device and at the other end we have the receiving device. In this chapter, we will use the Arduino board as the micro-controller board and the nRF24L01 transceiver chip as the RF transmitter as well as RF receiver. This concept is also known as **Machine-to-Machine (M2M)** communications: one microcontroller sends an RF encoded message to another microcontroller.

Security consideration:

 RF signals travel freely in the electromagnetic spectrum around us, in the air. The transmitted signals can be received by all RF receivers (apart from the intended recipient) in the vicinity of the transmitter. Hence the transmitted data can also be intercepted and received by anyone. In order to secure RF communications from unintended recipients, usually, RF signals are encrypted and they also require passwords.

For example, think of a home Wi-Fi router. A Wi-Fi router requires a password- without the password a connection to the router is not possible. But the Wi-Fi connection also warns us that "Information sent over this network might be visible to others". At the end of this chapter we will be able to appreciate why wireless data transfer is so vulnerable, unless protected.

Using the nRF24L01 transceiver module

The nRF24L01 RF transceiver is manufactured by Nordic semiconductor. It is a full feature radio frequency transceiver that has very low power consumption rating. For detailed specifications, you can visit the nRF24L01 product website hosted by **Nordic Semiconductor**. Some of the common devices that can be built are: wireless PC peripherals (mouse, keyboards, and headsets), remote controls for consumer electronic gadgets, and so on. However, it is not recommended for life support equipment development:

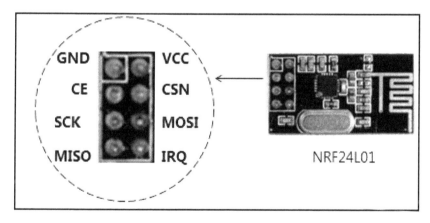

Figure 2: nRF24L01 pinout

The nRF24L01 transceiver module has eight pins available for connecting with external microcontroller devices. The chip itself is centrally located on the board shown as shown in the preceding figure. The eight pins are shown in the preceding pinout diagram. Additionally, there is an onboard voltage regulator to make the module tolerant to 5V input. Also notice the onboard antenna on the right-hand side. Some nRF24L01 boards come with the provision to attach an external antenna for increasing the range of the RF signals.

Wiring nRF24L01 with Arduino

In this section, we will learn how to wire the nRF24L01 with the Arduino board.

The following parts will be required for the transmitter circuit:

- One Arduino Uno R3
- One USB cable
- One nRF24L01 transceiver module
- Seven male-to-female jumper wires

The following parts will be required for the receiver circuit:

- One Arduino Uno R3
- One USB cable
- One nRF24L01 transceiver module
- Seven male-to-female jumper wires

We are going to wire up both the transmitter as well as the receiver circuits by following the common wiring scheme outlined in the following table:

Arduino Uno pin	nRF24L01 pin	Comments
3.3V	VCC	Connect directly
GND	GND	Connect directly
Digital pin 7	CSN	Connect directly
Digital pin 8	CE	Connect directly
Digital pin 11	MOSI	Connect directly
Digital pin 12	MISO	Connect directly
Digital pin 13	SCK	Connect directly

Arduino Uno pin	nRF24L01 pin	Comments
Not applicable	IRQ	Unused

Table 1: Arduino to nRF24L01 connections

Once all the parts are in place, go ahead and build the circuit as shown in the following diagram. Follow the same wiring scheme for both the transmitter as well as the receiver circuit.

At the end of this section, our aim will be to build two separate devices - one transmitter and one receiver. You can then power both the Arduinos from the USB of the same computer.

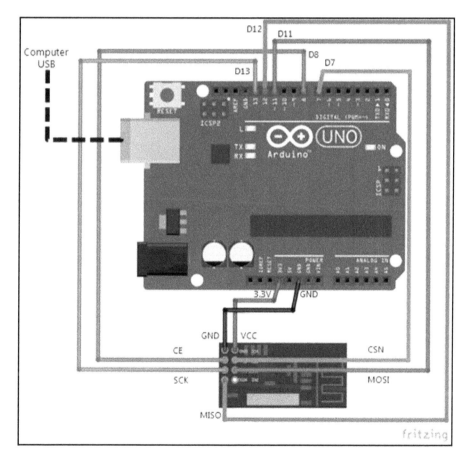

Figure 3: nRF24L01 to Arduino wiring

After building both the transmitter and receiver, load the following sketches. Load the RF transmitter sketch in the prototype that you want to use as a transmitter. Similarly, load the receiver sketch in the other prototype.

Downloading the open source RF library for Arduino

Before we can start writing the sketches, we must download and install the Arduino library for operating nRF24L01. The following are the steps to be followed:

1. Browse to GitHub for the **maniacbugRF library** using the URL:
   `https://github.com/maniacbug/RF24.`
2. Download the driver ZIP file from GitHub.

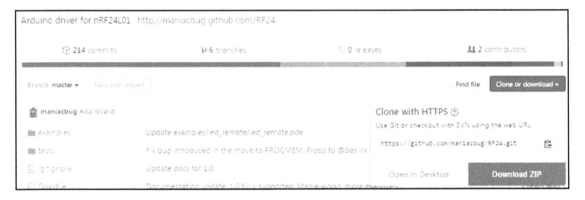

Figure 4: Download RF library

3. Unzip the downloaded zipped folder and place it in the Arduino libraries folder
   `C:\Program Files (x86)\Arduino\libraries:`

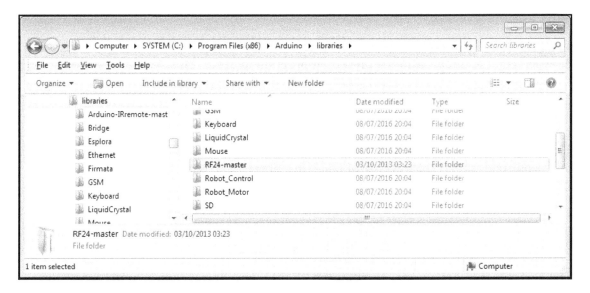

Figure 5: Paste unzipped library folder in Arduino libraries path

Transmitting radio frequency waves

After installing the RF library, we are now all set to start writing the sketches. In this section, we will learn how to transmit radio frequency waves. We will use the following C sketch for transmitting radio frequency waves as shown below :

```
#include <SPI.h>
#include "RF24.h"

int msg[1];// use this for transmitting integers
//char msg[1];// use this for transmitting characters

RF24 radio(8,7);
const uint64_t pipe = 0xE8E8F0F0E1LL;

void setup(void)
{
  radio.begin();
  radio.openWritingPipe(pipe);
}

void loop(void)
{
  msg[0] = 1;      // integer will be sent
  //msg[0] = 'b'; // character will be sent
```

```
    radio.write(msg, 1);
}
```

Now let us try to understand the preceding program in detail. This sketch uses two Arduino header files and includes them at the beginning of the sketch:

```
#include <SPI.h>
#include "RF24.h"
```

The `SPI.h` is the Arduino header file that is used for serial communications between microcontroller chips. The purpose of using this library is to write bytes onto the nRF24L01 chip from Arduino.

The second Arduino library, `RF24.h` contains all the low-level code that is necessary to transmit radio frequency waves. We are using the functions in this library to interface our Arduino sketch with the nRF24L01 chip for transmitting RF signals.

The next step is to declare the object for the RF chip. This line of code is used to create an object of type RF24. The RF24 type is defined in the Arduino library. The arguments 8 and 9 are Arduino's digital pins that will be connected to the nRF24L01 chip:

```
RF24 radio(8,7);
```

In the next line of code, we will declare an address which will serve as a communication pipe. This address space will be used as a buffer to store the data to be transmitted:

```
const uint64_t pipe = 0xE8E8F0F0E1LL;
```

The radio device object is initialized in the `setup()` function by beginning the communication channel and then issuing a command to open a pipe for writing data to the RF chip:

```
radio.begin();// start radio device object
radio.openWritingPipe(pipe);// open data transfer pipe
```

The main logic of the program is very simple and straightforward. First, we must declare a variable to hold a piece of information:

```
msg[0] = 1;//integer value 1 will be sent
```

In the preceding line of code, an integer type array variable is being initialized with the integer value `1` as the data. We are going to transmit this information via radio frequency by issuing the following write command to the serially attached nRF24L01 chip:

```
radio.write(msg, 1);// data gets transmitted via radio waves
```

Receiving radio frequency signals

In the previous section, we learnt how to transmit data via radio frequency waves. In this section, we will learn how to receive radio frequency waves. We will use the following C sketch for receiving radio frequency waves:

```
#include <SPI.h>
#include "RF24.h"

RF24 radio(8,7);
const uint64_t pipe = 0xE8E8F0F0E1LL;
int msg[1];        // use this to receive integers
//char msg[1];     // use this to receive characters

void setup()
{
  Serial.begin(9600);
  radio.begin();
  radio.openReadingPipe(1,pipe);
  radio.startListening();
}

void loop()
{
  if (radio.available())
  {
    bool endOfMsg = false;
    while (!endOfMsg)
    {
      endOfMsg = radio.read(msg, 1);
      Serial.println (msg[0]);
    }
  }
}
```

Now let us try to understand the preceding program in detail. This sketch uses two Arduino header files and includes them at the beginning of the sketch:

```
#include <SPI.h>
#include "RF24.h"
```

As explained in the previous section, the SPI.h is the Arduino header file that is used for serial communications between microcontroller chips. The purpose of using this library is to read bytes from the nRF24L01 chip attached with the Arduino.

The second Arduino library, RF24.h, contains all the low-level code that is necessary to receive radio frequency waves. We are using the functions in this library to interface our Arduino sketch with the nRF24L01 chip for receiving RF signals.

The next step is very similar to the previous example. Simply declare the object for the RF chip. The arguments 8 and 9 are Arduino's digital pins that will be connected to the nRF24L01 chip:

```
RF24 radio(8,7);          // 8 and 9 are Arduino's digital pins
```

Once again, we will declare an address which will serve as a communication pipe. This address space will be used as a buffer to store the data to be received:

```
const uint64_t pipe = 0xE8E8F0F0E1LL;
```

The radio device object is initialized in the setup() function by beginning the communication channel and then issuing a command to open a pipe for writing data to the RF chip. An extra line of code is required to start listening for ambient radio frequency signals:

```
radio.begin();// start radio device object
radio.openReadingPipe(1,pipe);// open data transfer pipe
radio.startListening();// start listening for RF signals
```

The main logic of the program is also straightforward. We must continuously check whether any RF signals are available by using the following line of code:

```
if (radio.available())
```

As soon as ambient radio waves are detected, the following program logic is used to read the message till the end:

```
bool endOfMsg = false;    // flag to indicate end of message
while (!endOfMsg)         // loop till end of message
{
  endOfMsg = radio.read(msg, 1);  // read the radio signal
  Serial.println(msg[0]);         // print message in serial monitor
}                                 // window
```

Thus, we can read RF signals by issuing the following read command to the serially attached nRF24L01 chip:

```
radio.read(msg, 1);// read received data via RF signals
```

Testing the RF transmitter-receiver pair

For testing the prototype, we will have to power up both the devices and connect a serial window to the receiver device prototype:

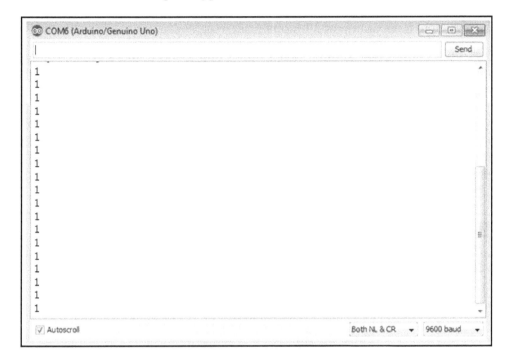

Figure 6: Testing transmitter prototype

Upon launching the serial monitor window for the transmitter, we will notice that the integer 1 will get displayed constantly on the screen.

Bluetooth communications

Bluetooth communications are also based on the 2.4 GHz ISM RF band. Bluetooth was originally invented and developed by Ericsson. The fundamental reason for inventing Bluetooth Technology was to provide a wireless alternative for RS-232 based cables and wires. Since then the Bluetooth protocol has come a long way and is primarily used over short distances (less than 30 feet) for:

- Establishing wireless connections between two hosts
- Exchanging data between two connected (paired) hosts

There are many varieties of Bluetooth chips available on the market today. Each variety is designed for meeting special purposes such as range, and special capabilities such as voice streaming. In this chapter, we will learn how to use Bluetooth communication by using the HC-05 Bluetooth chip. The HC-05 is a very commonly used Bluetooth chip with a synchronous transmission speed of 1 mbps and an Asynchronous transmission speed of 160 kbps. It can go upto a maximum of 2.1 mbps.

The HC-05 is a dual purpose module and can be configured as both a slave and a master. In Bluetooth communications we always create a pair of devices - one is a master and the other is a slave. There are some other varieties of the HC series Bluetooth module available in the market such as HC-06. The HC-06 can be used only as a slave only module. The HC-05 module is a classic Bluetooth module. There is another category of Bluetooth modules that consume very low energy, such as an HM-10 module, and are known as **Bluetooth Low Energy** (**BLE**) module. In this chapter, we will focus on the HC-05 which is based on the classic Bluetooth module.

Using the HC-05 Bluetooth module

The HC-05 Bluetooth module is a **Class 2** Bluetooth device. Class 2 means it has a range of roughly 30 feet. The HC-05 Bluetooth module is shown as follows:

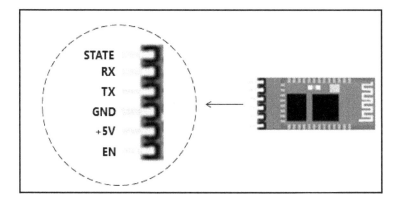

Figure 7: HC-05 Bluetooth module pinout

Connecting HC-05 to Arduino Uno

In this section, we will learn how to connect the HC-05 blue tooth module to the Arduino Uno. While doing so, we will learn a fundamental technique of reducing voltage on a particular line, known as the **voltage division** technique.

We will need the following parts for building this prototype:

- One Arduino Uno R3 + 1 USB A to USB B cable
- One breadboard
- One piece 1K Ohms resistor + 1 pc. 2K Ohms resistor
- Some male-to-male jumper cables

Let us wire up the HC-05 with the Arduino Uno by following the wiring scheme outlined in the following table:

Arduino Uno pin	HC-05 pin	Comments
5V	5V	Connect directly
GND	GND	Connect directly
Digital pin 0 (Rx)	TX	Connect directly
Digital pin 1 (Tx)	RX	Connect via voltage division technique

Table 2: Arduino to HC-05 connections

An important point to note is that the RX pin of the HC-05 will need an input voltage of 3.3 volts (higher voltages may damage the HC-05 chip). Since Arduino Uno's digital pins operate at 5 volts, we will have to use the voltage division technique to lower the 5 volts to 3.3 volts:

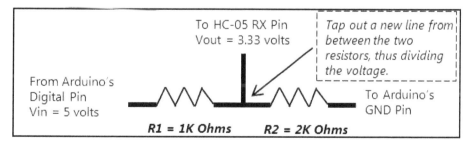

Figure 8: Voltage division technique

As per the voltage division formula: Vout = Vin x (R2 / (R1 + R2))

= 5 x (2 / (1 + 2)) => 3.33 volts

Follow the wiring diagram shown below for building the Bluetooth-enabled prototype based on the Arduino platform:

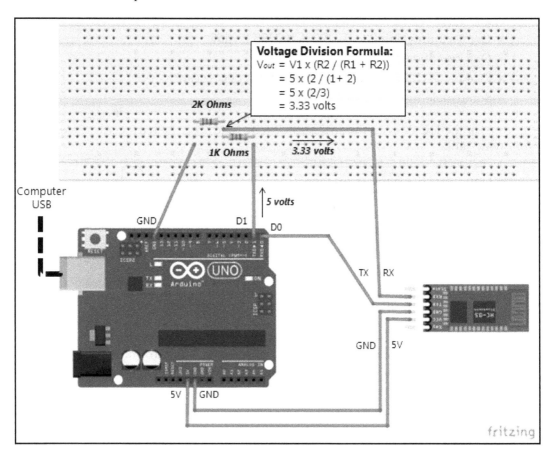

Figure 9: Wiring of HC-05 with Arduino Uno

HC-05 sketch

In this section, we will learn how to write a basic sketch to use the HC-05 Bluetooth module with Arduino.

Caution: Remove the jumper wires from Arduino's digital pins 0 and 1 before flashing the following sketch. Otherwise, the Arduino sketch upload will attempt to load the sketch into the HC-05 Bluetooth via the hardware serial pins (D0 and D1).

```
// include header files
String _inputString = "";
int _num = 0;

void setup()
{
  Serial.begin(9600);
  pinMode(13, OUTPUT);
}

void loop()
{
  if(Serial.available())
  {
    while(Serial.available())
    {
      //read the input from bluetooth serial
      char _ch = (char)Serial.read();
      _inputString += _ch;
    }

    _num = _inputString.toInt();
    if (_num > 0 && _num < 3)
    {
      // blink the onboard LED
      for (inti = 0; i< _num; i++)
      {
        digitalWrite(13, HIGH);
        delay(500);
        digitalWrite(13, LOW);
        delay(500);
      }
    }
  }

  _inputString = "";
  _num = 0;
}
```

The preceding sketch is designed to receive a message from another paired Bluetooth device. This sketch expects the incoming message to be an integer between 1 and 3. Based on the integer value received, Arduino Uno's onboard LED will be flashed the same number of times as the value of the integer received.

Communicating with the HC-05 prototype

Now that our Bluetooth prototype is up and running, the next step is to connect to it and send it some data. In this chapter, we will use the Bluetooth terminal client software from a Microsoft Windows mobile phone. However, you can do the same from any Bluetooth client, be it an iPhone, an Android, another laptop/PC, or any other Bluetooth chip connected to a micro-controller.

To communicate with your HC-05 prototype, perform the following steps:

1. Simply install Bluetooth terminal software on a device of your choice.
2. Launch the Bluetooth Terminal; you will see a list of Bluetooth devices in your vicinity.
3. Find the HC-05 in the list.
4. Follow the normal process of tapping on the listed device and then pair it by following the instructions on your Bluetooth terminal software. If prompted for a password then the most likely default factory set password for the HC-05 module would be either 1234 or 0000 (this would also be mentioned in the HC-05 datasheet):

Figure 10: Talking to HC-05 from Bluetooth terminal on a Windows phone

5. Once the HC-05 chip is paired with your device, taking the example of a Windows Phone shown, simply type in an integer number and send it. Simultaneously, you will notice that the onboard LED starts glowing ON and OFF.

Try the following

Let us try the following things before proceeding to the next chapter.

- In the nRF24L01 example, try to send a character instead of an integer
- In the HC-05 example, add a Piezo Buzzer to the Arduino board and beep it instead of flashing the onboard LED
- Extend this knowledge to communicate between two Arduino boards, each having their own HC-05 modules

Things to remember

Remember the following important points about what we learnt in this chapter.

- The 2.4 GHs ISM band is available for use by the industrial, scientific, and medical community
- Use of RF communications is strictly regulated by law
- RF signals can be received by all RF receivers in the vicinity
- Sensitive RF communications must be secured using encryption, public-private tokens, and so on
- Bluetooth is a wireless alternative to wired communications
- Voltage division technique is used to reduce voltage

Summary

In this chapter, we learnt how to build a prototype for interfacing RF devices with the Arduino board. In the first part of the chapter, we learnt how to use the nRF24L01 chip with the Arduino. Now you can employ the RF data transmission techniques learnt in this chapter to build the remote controlled reconnaissance boat mentioned in `Chapter 7`, *Day 5 - Using Actuators*.

In the second part, we used the HC-05 Bluetooth chip and learnt how to communicate using the HC-05 module with another Bluetooth enabled device (a mobile phone). You can extend this knowledge to communicating between any two machines/gadgets, each having their own HC-05 modules.

With this we shall conclude our learning of short range wireless communications on the Arduino platform. You can utilize the basics learnt in this chapter and attempt more challenging projects. In the next chapter, we will learn how to use a GSM module for performing basic telephony (SMS and phone calls) over long distances.

11

Day 9 - Long-Range Wireless Communications

"Mr. Watson - come here - I want to see you."
- Alexander Graham Bell

This chapter will be a fascinating introduction to the exciting, and sometimes secret world of telephony. Yes, we have all used it, but today we will learn about it! Looking back in time, remember the first day of this ten-day journey. I must say, you have done extremely well to have reached up to this point. So, let us buckle up and get started on yet another adventurous day with our favorite Arduino.

In this chapter, we will learn how to use a **GSM module** to dial phone numbers and send SMSs. The GSM module is very handy for long range communication in remote places that do not have Wi-Fi connectivity, but have telephone network coverage. The chapter will explain how a sample forest fire early warning system could be built using a simple GSM module.

You will learn about the following topics in this chapter:

- The GSM module
- Basic AT commands
- GSM module interfacing with Arduino
- GSM module sketch for sending SMS messages
- GSM module sketch for dialing phone numbers
- Forest fire early warning system project idea

The GSM module

Microcontroller-based telephony has a myriad of applications in the real world, from the banks and financial institutions sending out tons of alert SMS messages to their customers, to advanced smart city travel systems sending out traffic updates via SMS to their subscribers. GSM/GPRS is a very popular method of implementing a field solution, be it for telephony or for the **Internet of Things** (**IoT**).

Global System for Mobile (**GSM**) communications is based on **second generation** (**2G**) telecommunications protocols, whereas **General Packet Radio Services** (**GPRS**) is primarily used for mobile data service over 2G and 3G telecommunications protocols.

The following figure shows a typical GSM module that can be interfaced and used with the Arduino platform:

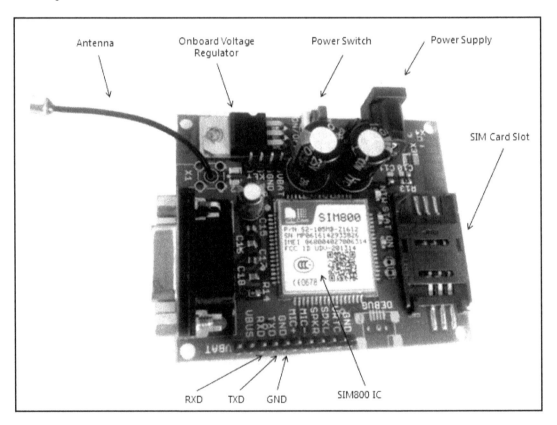

Figure 1: A Typical GSM module

As shown in the preceding figure, the GSM module is an integrated electronic board that houses multiple electronic components that are necessary for achieving the actual act to establishing telephony. These components include the following:

- The SIM800 GSM/GPRS chip
- A SIM card holder slot
- An active SIM card
- The GSM module antenna
- Various header pins for interfacing with external devices (such as an Arduino)

The GSM module used for illustration of the preceding components was procured from Elementz Engineers Guild Pvt. Ltd.

The most important part of the GSM module is the SIM chip. In the preceding figure, you will notice the central chip SIM800. This chip contains all the necessary telephony software. By sending AT commands, we can instruct this chip to dial numbers and send SMSs.

The SIM800 chip is manufactured by SIMCom. SIMCom is a company that specializes in the manufacturing of various types of telecom chips. An earlier version of the SIM800 module is the SIM900 module. The SIM900 module used to be widely used with the Arduino platform. However, the SIM800 is the latest version of the GSM chip.

The SIM800 is actually a quad-band GSM/GPRS chip that can be used for telephony in the following four frequency bands:

- 850 MHz
- 900 MHz
- 1800 MHz
- 1900 MHz

The SIM800 chip is fully loaded and also comes with Bluetooth and various other useful features. However, in this chapter we will concentrate on using a SIM800 based GSM module with the Arduino platform. Another option is using Arduino's stackable GSM shield that can be vertically stacked on top of the Arduino Uno main board. You can find more information on the stackable GSM shield on the official Arduino foundation website at: `https://www.arduino.cc/en/Main/ArduinoGSMShield`.

AT commands

AT commands are simple text commands that can be sent over serial connections to various electronic chips that conform to AT command instructions. Historically, the AT commands set is also known as the **Hayes command set**.

These commands were originally developed by Dennis Hayes. Dennis Hayes was the founder of Hayes microcomputer products that specialized in manufacturing modems. Since then, AT commands have become the popular choice among communication device manufacturers for sending commands to embedded communication software in the integrated chips:

AT command	Description
AT	This is the most basic AT command; its purpose is to ping a modem device. If the modem device is working correctly, then the response OK will be received from the modem.
AT+CMGF=1	This command is to set the SMS format to ASCII characters.
AT+CMGS="1234567890"	This AT command is used to instruct the modem to send an SMS. The number 1234567890 should be replaced by the phone number to which an SMS will be send.
ATD1234567890;	Use this command to instruct the modem to establish a voice call with a telephone. The telephone number 1234567890 shown in the command should be replaced with a phone number that will receive the call.
ATH	This command is used to disconnect the phone call session.

Table 1: AT commands used for basic telephony

The command set is characterized by short text strings, as shown in the preceding table, to send commands for operations such as dialing, hanging up, sending SMSs, reading SMSs, and so on. The preceding, listed AT commands have been used in this chapter. There are many AT commands in existence today. For a complete list of AT commands, you can search online.

GSM module interfacing with Arduino Uno

Compared to other modules that we have used so far, the GSM module is much larger in size. There are many pins that can be utilized for various purposes. However, for basic telephony, very few pins need to be connected with the Arduino:

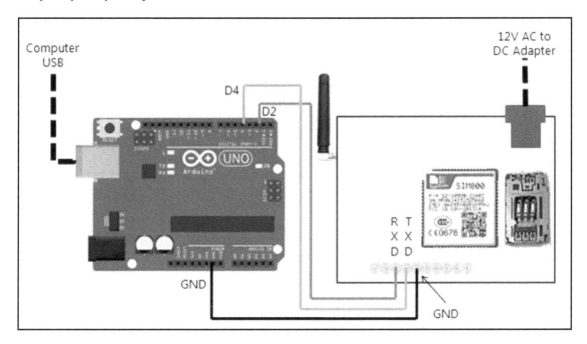

Table 2: GSM to Arduino wiring

Remember that you will find multiple GND pins on the GSM module board. If the incorrect GND pin is used, then your module will start getting overheated and will eventually burn out. For basic telephone use, the GND pin that is located beside the TXD and RXD pins.

If unsure of the correct GND pin, then there is an easy trial and error method to find out the correct GND pin. After powering up and connecting the module to the Arduino, if the SIM chip starts getting hot then you are most probably using the incorrect GND pin. The chip should not become hot if the correct GND pin is used.

Before loading the sketch, you must insert a normal SIM card into the GSM module's slot for the SIM card. After inserting the SIM card, switch on the GSM module.

Wait for 5-10 seconds or for some time until the **network** (**NTW**) LED on the GSM module starts blinking at regular intervals of 3-5 seconds. This regular blinking means the SIM card has registered itself on the telephone network and is ready to be used.

GSM module sketch

In this section, we will write the basic Arduino sketch for interacting with the GSM module. The following sketch has been written to dial a phone number as well as to send a text message via SMS.

The following sketch can be downloaded from the online location mentioned in the `Chapter 1`, *Boot Camp* of this book:

```
// Include the software serial library
// Alternately you can directly use the hardware serial
// via (Arduino's digital pins D0 and D1)
#include <SoftwareSerial.h>

// Define the two pins that will be connected
// between the Arduino Uno and the GSM module

// D2 is connected to RXD pin of the GSM module
// So D2 will be simulated as the Tx pin on the Arduino
#define SIM800_RXD 2

// D4 is connected to TXD pin of the GSM module
// So D4 will be simulated as the Rx pin on the Arduino
#define SIM800_TXD 4

// Define the object for using the software serial
// these two pins are used to transmit the AT commands
// from the Arduino board to the modem in the GSM module
// SYNTAX: SoftwareSerialobj(Arduino_Rx, Arduino_Tx)
SoftwareSerial GSM_SIM800(SIM800_TXD, SIM800_RXD);

// The number 1234567890 shown below
// should be replaced with the number to be called
String phoneNumber = "1234567890";
```

```
//Initialize the sketch for using the GSM module
void setup()
{
  // initialize the hardware serial
  // for communication with the Serial Monitor window
  Serial.begin(9600);

  // initialize the software serial
  // for communication with the GSM modem
  GSM_SIM800.begin(9600);

  // Wait for 3 seconds for the SIM800 module to initialize
  delay(3000);
}

// Define the main logic of the main program loop
void loop()
{
  // read a character from the Serial Monitor window
  char ch = Serial.read();

  // if the character a is entered
  if (ch == 'a')
  {
    // invoke function to send an SMS
    sendAlertSMS();
  }

  // if the character b is entered
  if (ch == 'b')
  {
    // invoke the function to dial a phone number
    callNumber();
  }
}

// Logic to send SMS
void sendAlertSMS()
{
  Serial.println("Sending SMS...");

  // Issue AT command to set SMS format to ASCII characters
  GSM_SIM800.print("AT+CMGF=1\r\n");
  delay(1000);

  // Issue an AT command to send a new SMS to a phone number
  GSM_SIM800.print("AT+CMGS=\"" + phoneNumber + "\"\r\n");
  delay(1000);
```

```
    // Next step is to send SMS content to the modem
    // Note that this step does not use a separate AT command
    GSM_SIM800.print("This is an SMS");
    delay(1000);

    // The last AT command is to
    // send Ctrl+Z /ESC character sequence
    // to indicate to the GSM modem
    // that the SMS message is complete
    GSM_SIM800.print((char)26);
    delay(2000);

    // Get results of sending SMS
    // Print the results in the Serial Monitor window
    if(GSM_SIM800.available())
    {
      Serial.println(GSM_SIM800.read());
    }
  }

// Logic to call phone number
void callNumber()
{
  // Send AT command TD to the GSM modem
  // along with the number to be called
  Serial.println("Calling...");
  GSM_SIM800.print("ATD" + phoneNumber + ";\r\n" );

  // wait 10 seconds before the next loop
  delay(15000);

  // Issue an AT command to the GSM modem to disconnect the call
  GSM_SIM800.print("ATH\r\n");
}
```

The preceding sketch is self-explanatory with in-line comments that explain what is going on. Hence, we will quickly look at the main features of the sketch. In the preceding sketch, we have used the software serial method. We already know that the Arduino board has two digital pins **D0 (Rx)** and **D1 (Tx)**. These two pins are internally used by Arduino whenever the following operations are happening:

- A sketch is being uploaded from a computer into Arduino's program memory
- We interact with the running Arduino sketch via the serial monitor window

- If external hardware is connected to **D0 (Rx)** and **D1 (Tx)**, and we execute the `Serial.print()` or `Serial.write()` function, then the data is written on pin **D1 (Tx)** - **Tx** indicates transmission pin.
- If an external hardware is connected to **D0 (Rx)** and **D1 (Tx)**, and we execute the `Serial.read()` function, then the data is read from pin **D0 (Rx)** - **Rx** indicates reception pin.

Recall that in the GSM wiring, we used digital pins **D2** and **D4** to communicate with the GSM modem. **D2** is connected to the **RXD** pin of the GSM, hence **D2** is simulated (by functionality in `SoftwareSerial.h`) as the **Tx** pin of the Arduino board. Similarly, **D4** is simulated as Arduino's **Rx** pin and is connected to the GSM module's **TXD** pin.

```
// SYNTAX: SoftwareSerialobj(Arduino_Rx, Arduino_Tx)
SoftwareSerial GSM_SIM800(SIM800_TXD, SIM800_RXD);
```

If we had wired Arduino's digital pins **D0 (Rx)** and **D1 (Tx)** to the GSM module then we would not have required the software serial.

Fundamental:

If more than one serial connection is required in a single setup, then implementing the software serial technique is useful. For example, in this case we need to use the serial monitor window (internally programmed to work with the hardware serial pins) to interact with the running sketch. At the same time, we need another serial connection for communicating with a peripheral device (the GSM module in this case). Thus, we must use the software serial technique in order to achieve an extra (simulated) pair of serial pins.

Now that we understand the concept of software serial, let us quickly browse through the sketch. The sketch starts with the customary sections followed by defining two fundamental functions:

- `sendAlertSMS()`
- `callNumber()`

The `sendAlertSMS()` function contains the logic for sending an SMS text message to a specified mobile number. It contains the series of AT commands used for sending an SMS by using the attached GSM module. The most important fundamental to note here is the usage of the `print()` method of the software serial:

```
GSM_SIM800.print("AT+CMGF=1\r\n");
```

This is how the AT commands are sent from the Arduino sketch to the attached GSM module.

It should be noted that if you try to use the `print()` method then the AT commands must be ended with `\r\n` characters to indicate the required carriage return and new line as per the requirements of the AT command. There are some cases where the `\r\n` characters are not required; in such cases only the `print()` command without the `\r\n` suffix will be sufficient. However, when using the `println()` method the `\r\n` characters are not required explicitly.

The next function, `callNumber()`, contains the logic for dialing a given mobile number from the Arduino sketch via the attached GSM module. This function contains the straightforward code to call a specified phone number. However, a note regarding the AT command being used to dial a phone number is that we must use a semi-colon (`;`) symbol at the end of the AT command:

```
GSM_SIM800.print("ATD" + phoneNumber + ";\r\n" );
```

The `phoneNumber` variable has been set at the top section of the sketch with the 10 digit mobile phone number. You must manually replace the digits `1234567890` in the sketch with the actual phone number at your end; there's no need to add the ISD code in this sketch. After string concatenation, the preceding code will look as follows:

```
GSM_SIM800.print("ATD1234567890;\r\n" );
```

The main `loop()` of the sketch is designed to continuously look for specific characters from the serial monitor window. Depending upon the character entered from the serial monitor window, either the function to dial a number or to send an SMS will be invoked.

If the lowercase letter `a` is entered, then the sketch will invoke the function to send an SMS to a specified phone number, as shown in the following code:

```
// if the character 'a' is entered
if (ch == 'a')
{
  // invoke function to send an SMS
  sendAlertSMS();
}
```

If the lowercase letter b is entered, then the sketch will invoke the function to dial a specified phone number, as shown in the following code:

```
// if the character 'b' is entered
if (ch == 'b')
{
  // invoke the function to dial a phone number
  callNumber();
}
```

Forest fire early warning system - Inspiration

Now that we know how to achieve basic telephony with the Arduino and a GSM module sketch, just imagine all the marvelous things that you can start making. Let us explore an exciting project idea using a basic GSM module.

In this section, we will put together the idea of an advanced standalone device for building an early warning system for wild forest fires. Getting notified of a forest fire in time can save many precious lives and natural resources:

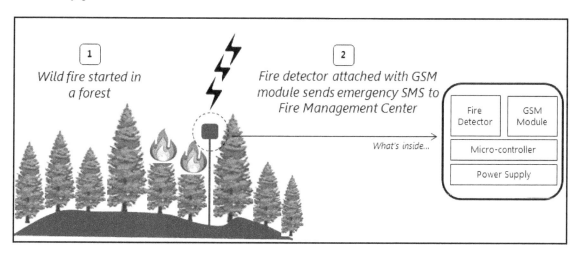

Figure 2: Early warning system for forest fires

As depicted in the preceding figure, a standalone device composed of the following units when strategically located at regular intervals in a forest can provide valuable early warning:

- A microcontroller (for prototyping, we will use the Arduino Uno)
- A suitable power supply

- A fire detector module
- A GSM module
- A weather proof container

For the sketch, we can reuse the function for sending SMS messages from this chapter. You can then leverage the technique learnt in Chapter 5, *Day 3 - Building a Compound Device* (for using a smoke detector) and apply the same to interface with a professional quality certified fire detector. The knowledge of building a standalone device (acquired in Chapter 6, *Day 4 - Building a Standalone Device*) will help you to calculate and choose a suitable power supply and outer packaging.

The following is the skeleton of a sketch that you can use for building the early warning system for wild forest fires:

```
//*********************************************************/
// Step-1: INITIALIZE LOCAL VARIABLES
//*********************************************************/
#include <SoftwareSerial.h>
// INSTRUCTIONS FOR WRITING ADDITIONAL BELOW CODE
// include Fire Detector library if required

#define SIM800_RXD 2
#define SIM800_TXD 4
SoftwareSerial GSM_SIM800(SIM800_TXD, SIM800_RXD);
// INSTRUCTIONS FOR WRITING ADDITIONAL BELOW CODE
// Define Fire Detector object if required

// The number 1234567890 shown below
// should be replaced with the number to be called
String phoneNumber = "1234567890";

//*********************************************************/
// Step-2: INITIALIZE I/O PARAMETERS
//*********************************************************/
void setup()
{
  GSM_SIM800.begin(9600);
  delay(3000);
  // INSTRUCTIONS FOR WRITING ADDITIONAL BELOW CODE
  // Refer to fire detector documentation
  // and initialize the Fire Detector if required
}

//*********************************************************/
// Step-3: MAIN PROGRAM
//*********************************************************/
```

```
void loop()
{
  // INSTRUCTIONS FOR WRITING ADDITIONAL BELOW CODE
  // Write a new function named readFireDetector()
  // Hint: Usually the logic is very simple and easily available
  // from the Fire Detector's documentation
  // Most often it would be reading the value of a digital
  // or analog pin using digitalRead() or analogRead() functions
  bool isFire = readFireDetector();

  // if a fire has been detected
  if (isFire)
  {
    // invoke function to send an SMS
    // REUSE the logic from this chapter
    // uncomment the below line after copying the function
    // from the GSM reference sketch
    // sendAlertSMS();
  }
}


bool readFireDetector()
{
  // Hint: Usually the logic is very simple and easily available
  // from the Fire Detector's documentation
  // Most often it would be reading the value of a digital
  // or analog pin using digitalRead() or analogRead() functions

  // Uncomment the following sample lines during implementation
  // Assuming the fire detector output is connected to D5
  // --------------------------------------------------------
  int fireDetectorOutputPin = 5;
  bool isFire = false;
  pinMode(fireDetectorOutputPin, INPUT);
  isFire = digitalRead(fireDetectorOutputPin);
  //return isFire;   // uncomment this line
  return false;      // comment this line
}
```

As explained in the embedded comments in the preceding sketch, you will have to implement a new function called readFireDetector(). This function should implement the logic to read the value of a fire detector sensor module.

Use the knowledge that you acquired in `Chapter 4`, *Day 2 - Interfacing with Sensors* and `Chapter 5`, *Day 3 - Building a Compound Device* in order to understand and write the Arduino sketch code for reading a value on a fire detector sensor's output pin.

In the next and last chapter of this book we will learn how to use the Arduino platform for building IoT projects.

Try the following

Let us try the following things before proceeding to the next chapter:

- Add an **SD card module** with the GSM setup. Read the SMS content from the SD card, instead of hard coding it in the sketch. You can utilize the knowledge gained from `Chapter 5`, *Day 3 - Building a Compound Device* for using the SD card module.
- Add an **HC-SR04 ultrasonic sensor** to the GSM setup. Keep the device in your backyard, pointing the ultrasonic sensor away from your house. Write a sketch to send you an SMS every time someone comes within 10 feet of the device. Make it call your phone if someone comes within five feet. You can reuse the knowledge of using the HC-SR04 from `Chapter 6`, *Day 4 - Building a Standalone Device*.

Things to remember

Remember the following important points about what we learnt in this chapter:

- AT commands, also known as the Hayes command set, are simple text-based commands that are used to instruct and control modem-based communications systems.
- The GSM module has multiple GND pins. Always connect the appropriate pin. Connecting a wrong GND pin will result in overheating and damage.
- Most GSM modules are rated between 1-2 Amps. However, 1.5 Amps power supply is often fine for normal use.
- After inserting a SIM card into the GSM module and powering it up, wait until the NTW LED starts blinking at regular intervals of 3-5 seconds.

Summary

In this chapter, we were introduced to the world of telephony. We looked at a basic telephony device, the SIM800-based GSM module. Then we learned how to write an Arduino sketch to send an SMS and dial a phone number from the Arduino.

In the last part of the chapter, we put together the idea of building an early warning system for wild forest fires. In this inspirational section, we realized that we can reuse our knowledge from the previous chapters and successfully build new projects. This is also what this book aimed to do, to empower you and make you conversant with creating new projects on the Arduino platform.

12
Day 10 - The Internet of Things

"Because the people who are crazy enough to think they can change the world are the ones who do."
- Steve Jobs

In this concluding chapter, we will learn how to use the Arduino platform in the fast emerging Internet of Things world. All of us have heard about the buzzword **Internet of Things (IOT)**. The IOT is a growing network of physical devices that can connect to the existing internet and exchange data with other devices.

For example, it is possible for the microprocessors embedded in a car in Paris, or in a car in New York, to communicate with another car parked in New Delhi--all this over the existing internet, but using some special platforms, techniques, and protocols. In this chapter, we will look at the basics of connecting physical devices to the internet and a simple technique of transmitting data from physical devices, irrespective of their location around the globe.

You will learn the following topics in this chapter:

- IoT concept
- IoT edge devices
- IoT cloud platforms
- Connectivity with ESP8266 Wi-Fi chip
- Smart retail project inspiration

Introduction to IOT

As per Gartner, the number of connected devices around the world is going to reach 50 billion by the year 2020. Just imagine the magnitude and scale of the hyper-connectedness that is being forged every moment, as we read through this exciting chapter.

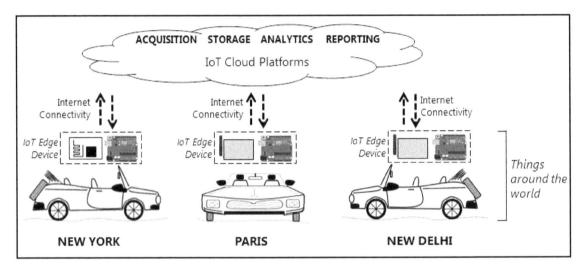

Figure 1: A typical IoT scenario (automobile example)

As we can see in the preceding figure, a typical IoT-based scenario is composed of the following fundamental building blocks:

- IoT edge device
- IoT cloud platform

An IoT device is used to serve as a bridge between existing machines on the ground and an IoT cloud platform.

The IoT cloud platform provides a cloud-based infrastructure backbone for data acquisition, data storage, and computing power for data analytics and reporting.

The Arduino platform can be effectively used for prototyping IoT devices for almost any IoT solution very rapidly.

IoT edge devices

The IoT edge devices help in connecting the existing machines to an IoT cloud platform. They are essentially hardware components with embedded software that provide an operating environment for running custom programs for:

- **Interfacing with connected machines**: The edge device sits in between the existing machines and the IoT cloud. It serves as a bridge over which data is exchanged between the existing machines and the IoT cloud.

- **Reading parameters or signals from connected machines**: The edge device also has custom code (for example, Arduino sketches) embedded and running inside it. The embedded sketch/programs serve to interface with the machine that is being connected to the internet.

- **Connecting to the internet**: The embedded programs in the edge devices help to connect easily to the existing internet. For this purpose, various types of connecting media are utilized. Some of the popular methods are wireless GPRS (General Packet Radio Service) or GSM (Global System for Mobile), Wi-Fi Chips, and hard-wired LAN cable shields.

- **Transmitting data to an IoT cloud**: The edge device also runs embedded software that provides functionality to send data to the IoT cloud platform being used. Often, the IoT cloud platform vendors provide their own device compatible header files, the use of which becomes very easy to establish connectivity and transmit data to the IoT cloud. It is also sometimes possible that the IoT device manufacturers themselves provide IoT cloud compatible header files that can be readily used with our programs/sketches. It depends on a case-to-case basis.

- **Fetching data from an IoT cloud**: Similarly, the edge device also runs embedded software that provides functionality to read data from IoT cloud platforms.

- **Hosting local endpoints for receiving requests from an IoT cloud**: This is an advanced technique; hence it is good to know the basic purpose of this technique. Often this technique is employed for edge devices to be invoked directly from the IoT cloud platforms using certain IoT protocols such as the **Messaging Queue Telemetry Transport** (**MQTT**) protocol.

- **Sending signals and controlling the connected machines**: The edge devices are also responsible for sending signals and controlling the physical machine that is being connected to the internet:

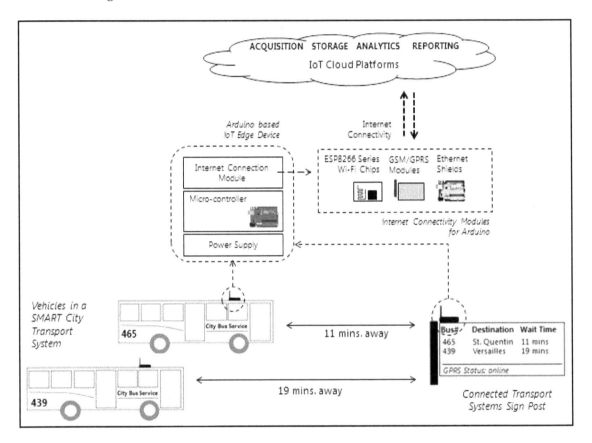

Figure 2: Atypical Arduino based IoT edge device prototype

In the preceding diagram, a very practical prototype of a smart city transport system has been depicted. The buses plying in the city continuously send data regarding their current location to the IoT cloud platform.

In each bus stop, there is an electronic sign post, which continuously fetches information regarding how far away a bus is and then displays the retrieved information in the information panel. Thus, the preceding IoT application fulfils a very important purpose for the citizens of the smart city.

As depicted earlier, a basic Arduino-based IoT edge device is composed of the following fundamental parts:

- A microprocessor or microcontroller unit (which in the case of our study will be the Arduino) that serves as the brain of the edge device. In actual practice, field grade microprocessor units are used for industrializing the actual edge device product.
- An internet connectivity module for facilitating internet connections. Some popular varieties include the GPRS/GSM module, Wi-Fi chips, and Ethernet shields. For learning the basics, we will use the ESP8266 Wi-Fi module with the Arduino platform.

IoT Cloud platforms

IoT cloud platforms are specialized cloud-based private or public infrastructure, which in itself is an extensive topic. We will focus on the very basics of IoT cloud platforms in this chapter. The IoT cloud platforms are carefully designed for use by edge devices. The IoT platforms usually provide the following fundamental services:

- Web-based administrative console for managing the connection details of edge devices. Device-specific connection points are also known as channels, while there are also generalized messaging queues to which any device can post data.
- Cloud-based backend databases/stores/lakes for storing the incoming device data.
- Cloud-based analytics services to operate upon the incoming data (also known as stream analytics) and transform the data into more meaningful desirable forms.
- To cap it all, some advanced reporting functionality is also provided by the cloud platforms.

There are many IoT cloud platform providers in the market, such as the following, to name a few. We must choose depending on the situation and the project:

- Azure IoT Hub/Event Hub from Microsoft
- AWS IoT from Amazon
- Google IoT
- IBM BlueMix IoT
- IBM Watson IoT
- ThingSpeak IoT
- Thingworx IoT

- Particle IoT
- Dweet IoT

In this chapter, we will use an IoT cloud platform known as ThingSpeak. The ThingSpeak platform has been chosen for ease of use with Arduino, easy availability of many online resources, and it also does not require a credit card for free registration and academic use.

 The cloud platform is officially available for registration at `https://thingspeak.com`.

The following diagram shows the basic general steps involved in IoT prototyping:

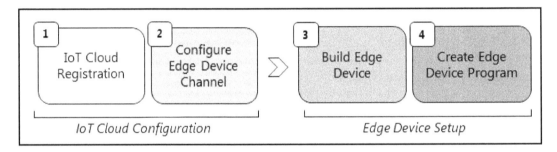

Figure 3: Basic steps in IoT prototyping

As shown earlier, the first step in IoT prototyping is to choose and register on a preferred IoT cloud platform. Once the basic registration has been completed, the next step is to set up the edge devices in the IoT cloud. For setting up a device in the IoT cloud, you will usually find a dashboard after you log in to your IoT cloud account. From this dashboard, there will be options to set up the details of the edge devices from which data will be transmitted.

The final part of the IoT prototyping process consists of two basic steps: building the actual edge device and then writing a program/sketch to be embedded in the edge device.

In the following sections, we will learn how to configure an IoT cloud and build the edge device using the Arduino board with the ThingSpeak IoT cloud platform via the ESP8266-01 Wi-Fi chip.

IoT cloud configuration

In this section, we will learn how to start using an IoT cloud platform. As already mentioned, there are many players in this space. The choice of IoT cloud will depend on your requirements. Let's move ahead and see how to register and then set up an IoT edge device to send data to the ThingSpeak IoT cloud platform.

Step 1 - IoT cloud registration

Registering on various available IoT platforms will be slightly different from one another; however, the basic steps would remain the same across the board. Let's follow the steps outlined here to register on the ThingSpeak IoT platform:

1. The first step is to create an account. While this book was being written, ThingSpeak did not require a credit card for signing up.
2. Sign-up at `https://thingspeak.com/users/sign_up`.
3. After signing up on ThingSpeak, go ahead and log in.
4. Login at `https://thingspeak.com/login`.

Once the preceding steps are completed, the next thing that we will have to perform is to start setting up our first connected device in the ThingSpeak IoT cloud. The process has been explained next in great detail.

But keep in mind that these steps are specific to a particular IoT platform. Therefore, when using another platform, you must adapt to the procedure defined by the other platforms. Usually, you can find the process of configuring edge devices in the IoT platform's documentation.

Step 2 - Configuring an edge device channel

The second step is to create a new device channel in the IoT cloud. A device channel is a defined endpoint (specific to a particular edge device) in the IoT cloud where we can transmit data by simply invoking URLs via HTTP. So, if we have 10 edge devices, then we will have to create 10 separate device channels in the IoT cloud.

In this section, we will learn how to create a ThingSpeak channel. Start by clicking on the **New Channel** button as shown in the following screenshot:

Figure 4: Creating a new channel

After clicking on the **New Channel** button, a new data entry screen will appear. In the new channel creation screen, fill out the details of the device from which the data will be received. For the Arduino sketch example provided in this chapter, you will just have to fill out the textboxes **Name**, **Description**, and **Field 1**, as shown in the following screenshot:

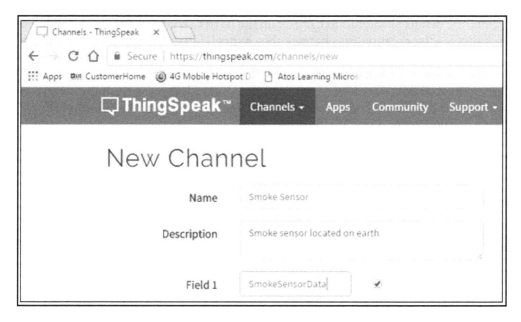

Figure 5: Setting up a new device channel

In the new channel creation screen, scroll down further and make this channel public (as shown next) by checking the **Make Public** checkbox, so that you can post data to it from the edge device:

Figure 6: Making the device channel publicly accessible

After filling the details as mentioned, check on the Save Channel button to create the new device channel. After creating the channel, the channel details screen will get displayed (as shown next). Notice that there will be five tabs: **Private View**, **Public View**, **Channel Settings**, **API Keys**, and **Data Import/Export**. That is all. You are now all set to start using the edge device and post data to the ThingSpeak IoT cloud. Browse through the five tabs to get a feeling of the contents under each of these tabs. Make note of the **Channel ID**, as you will need it for posting and reading data to ThingSpeak:

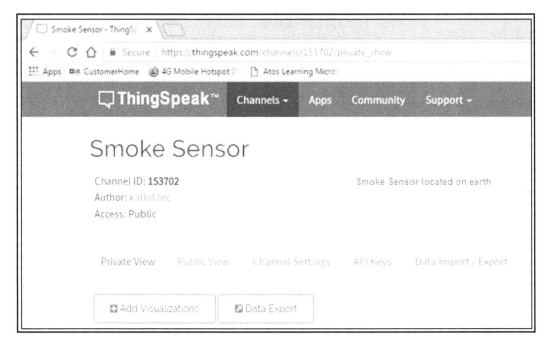

Figure 7: Channel details dashboard

Now we will see how to log data and read data to and from this channel. Browse to the **API Keys** tab for your channel:

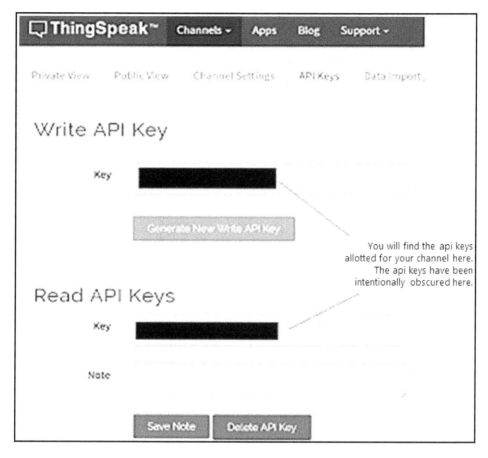

Figure 8: Channel API keys

Note down the **Read API Keys** and **Write API Key** (as shown earlier), as you will need it for reading and posting data from and to the ThingSpeak IoT cloud respectively. The API keys for the channel shown in the preceding figure have been obscured intentionally.

 Never share your channel API keys in order to secure your device channel from unauthorized or inadvertent spamming.

Now, let's understand how to send data to the newly created device channel. First we will learn how to simply use an URL and hit the URL in a web browser, and send the desired data to the ThingSpeak IoT cloud platform. Once we have understood how the URL was used, then we will attempt the same thing from an Arduino sketch. Thus, we will effectively learn how to send data from an edge device to an IoT cloud.

You can use the following URL format to log data to ThingSpeak:
`https://api.thingspeak.com/update.json?api_key=<your-write-api-k ey>&field1=58.`

In the preceding URL format, replace `<your-write-api-key>` with the API Write Key for your device channel. The `field1=58` refers to the `field1` that was set up while creating the device channel. You should replace the value `58` by the actual value that has to be sent to the cloud:

Figure 9: Recording data into ThingSpeak via the URL in the browser

Upon hitting the URL in a web browser, the data '58' will get logged in **field1** for the configured device channel in the ThingSpeak IoT cloud. Immediately after the URL is hit, you will see the preceding response in the browser. The response text is the JSON (JavaScript Object Notation) format response sent from the ThingSpeak device channel endpoint. If you inspect the response displayed in the browser, then it can be seen that it represents a data row from a table that got populated with some meaningful data when we hit the URL in the browser. In other words, when the URL was hit in the browser, the data was sent to the device channel endpoint and got stored in an internal database table corresponding to the device channel. So every time we want to send a new data point to the cloud, the same URL format can be used. The only change would be the value to be posted.

Similarly, we can utilize the same URL format to send an HTTP get request from an Arduino sketch. However, the Arduino board will require an additional piece of hardware for connecting to the internet. As already explained at the beginning of this chapter, there are various hardware options available for connecting to the internet. In this chapter, we will learn how to use the ESP8266-01 Wi-Fi chip with Arduino Uno.

After sending data to the ThingSpeak IoT platform, the next step in understanding the process is to browse to the web dashboard from where the newly logged data can be viewed in a graphical format. In case of the ThingSpeak cloud, you can also see the logged data in the ThingSpeak channel's **Private View** and **Public View** tabs, as shown in the following screenshot:

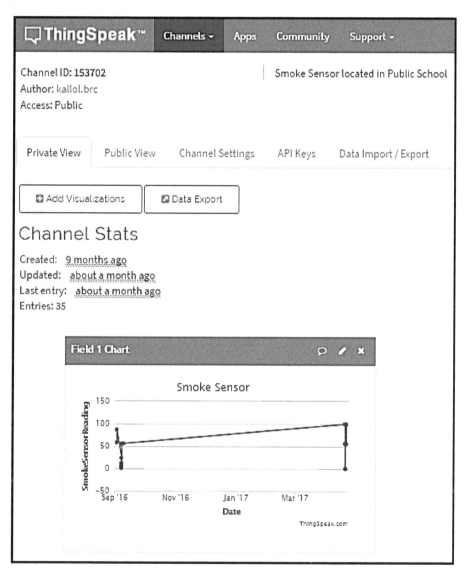

Figure 10: Viewing logged data

Now that we have understood how to send data to an IoT cloud by using very simple URL formats, let's learn the technique to fetch data. Once again, we will use very simple URL formats for fetching data from the ThingSpeak IoT cloud.

The important point to note here is the structure and/or format of the URL and the protocols used that vary across various IoT cloud platforms. Depending on the platform being used, you should refer to the platform-specific documentation.

 Use the following URL format to read the logged data from ThingSpeak:
`http://api.thingspeak.com/channels/<your-channel-id>/feeds.json?key=<your-read-api-key>&results=1`.

On hitting the preceding URL in the browser, you will receive the last posted data row from ThingSpeak in the following screenshot:

```
{"channel":{"id":153702,"name":"Smoke Sensor","description":"Smoke
Sensor located in Public
School","latitude":"0.0","longitude":"0.0","field1":"SmokeSensorRead
ing","created_at":"2016-09-02T17:12:06Z","updated_at":"2017-06-
12T17:35:08Z","last_entry_id":36},"feeds":[{"created_at":"2017-06-
12T17:35:09Z","entry_id":36,"field1":"58"}]}
```

Figure 11: Fetching the logged data from ThingSpeak in the browser

Using JSON-parsing techniques, we can extract the value for **field1**. Once the value has been extracted, it is up to the consuming application (either in the cloud or on a device) to decide what needs to be done next. JSON (JavaScript Object Notation) is a web development technology used to represent JavaScript objects in plain text format. Since JSON is a separate topic and is beyond the scope of this book, you can use the following online resource for learning the basics of JSON:

`https://www.w3schools.com/js/js_json_intro.asp`

Basically, you are free to use any technology of your choice that you are comfortable with, to parse the received data from the cloud. However, JSON has been mentioned here, because the data returned by the cloud is readily available in JSON format, as shown in Figure 11.

Edge device setup

Now, let's start building our first basic IoT edge device. We will use the ESP8266-01 Wi-Fi capable chip. This chip comes loaded with networking layer software and can directly accept AT commands from the Arduino (or any serial terminal) for establishing internet connectivity with ambient Wi-Fi networks:

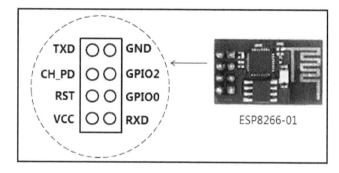

Figure 12: ESP8266 Wi-Fi chip pinout

The ESP8266-01 Wi-Fi transceiver module has eight pins available for connecting with external microcontroller devices. The eight pins are shown in the preceding pinout diagram. There is also an onboard antenna on the right-hand side. This chip can be used as a Wi-Fi connector from the Arduino board. It can also be used as a standalone node MCU (microcontroller unit) by directly loading a sketch into the ESP8266 board, without the Arduino Uno. However, in this chapter, our focus will be on the basic feature of connecting to the internet by using this chip.

Building the edge device

In this section, we will learn how to use the ESP8266 Wi-Fi chip with the Arduino Uno for connecting to the internet and posting data to an IoT cloud. There are numerous IoT cloud players in the market today, including Microsoft Azure and Amazon IoT. In this book, we will use the ThingSpeak IoT platform, which is very simple to use, with the Arduino platform.

The following parts will be required for this prototype:

- One Arduino Uno R3
- One USB cable
- One ESP8266-01 Wi-Fi transceiver module
- One breadboard
- One pc. 1K Ohms resistor
- One pc. 2K Ohms resistor
- Some jumper wires

Once all the parts have been assembled, follow the breadboard circuit shown in the following figure and build the edge device:

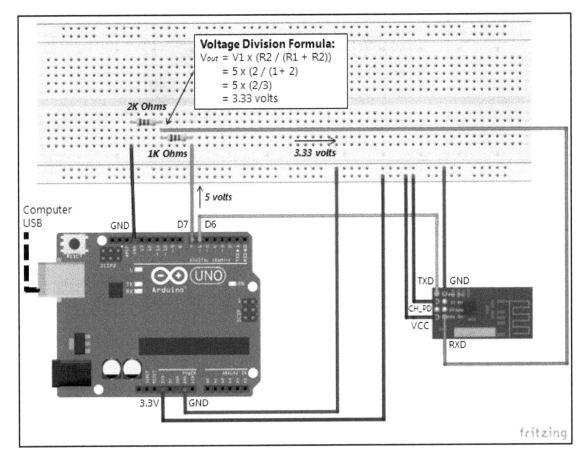

Figure 13: ESP8266 with Arduino Uno wiring

The important facts to remember in the preceding setup are:

- The RXD pin of the ESP8266 chip should receive a 3.3V input signal. We have ensured this by employing the voltage division method.
- For test purposes, the preceding setup should work fine. However, the ESP8266 chip is demanding when it comes to power (read current) consumption, especially during transmission cycles.

 Just in case the ESP8266 chip does not respond to the Arduino sketch or AT commands properly, then the power supply may not be enough. Try using a separate battery for the setup. When using a separate battery, remember to use a voltage regulator that steps down the voltage to 3.3 volts before supplying the ESP8266 chip. For prolonged usage, a separate battery based power supply is recommended.

Edge device sketch

Once you have built the edge device circuit as explained in the previous section, go ahead and load the following sketch for ESP8266 into the Arduino Uno board:

```
#include <SoftwareSerial.h>

//Usage: object-name (RX-pin, TX-pin)
SoftwareSerial ESP8266 (6, 7);

void setup()
{
  // Configure the hardware and software Serial
  Serial.begin(9600);
  ESP8266.begin(9600);
}

// main program logic to run repeatedly
void loop()
{
  // keep checking for user input from the Serial Monitor window
  char cmd = Serial.read();
  if (cmd == '1') { AT(); }
  if (cmd == '2') { connectWifi(); }
  if (cmd == '3') { CIPSTART(); }
  if (cmd == '4') { CIPSEND(); }
  if (cmd == '5') { sendRequest(); }
  if (cmd == '6') { disconnectWifi(); }
}
```

```
// logic to check whether the ESP8266 chip is working properly
void AT()
{
  Serial.write("AT");
  ESP8266.write("AT\r\n");
  delay(1000);
  if(ESP8266.find("OK")) { Serial.println("OK"); }
  else { Serial.println("Not Working"); }
}

// logic to connect to Wi-Fi router
void connectWifi()
{
  Serial.println("Connecting to Wi-Fi ...");

  // Replace the Wi-Fi SSID and PASSWORD shown below
  // with the SSID and PASSWORD of your Wi-Fi network
  ESP8266.write("AT+CWJAP=\"SSID\",\"PASSWORD\"\r\n");
  delay(7000);
  Serial.println(ESP8266.readString());
}

// logic to initiate a new transmission to ThingSpeak IoT cloud
void CIPSTART()
{
  // You can use the below IP address as-is
  // Because this is the ThingSpeak IP address and will not change
  ESP8266.write("AT+CIPSTART=\"TCP\",\"184.106.153.149\",80\r\n");
  delay(5000);
  Serial.println(ESP8266.readString());
}

// logic to specify data length to be transmitted
void CIPSEND()
{
  // The data length 46 was calculated for the request payload
  // GET /update?key=ID83C0GFDJM9B7GX&field1=56\r\n
  // This payload has been used in the function sendRequest()
  ESP8266.write("AT+CIPSEND=46\r\n");
  delay(3000);
  Serial.println(ESP8266.readString());
}

// logic to transmit data and signal end of transmission
void sendRequest()
{
  ESP8266.write("GET /update?key=ID83C0GFDJM9B7GX&field1=19\r\n");
  ESP8266.write("\r\n");
```

```
    delay(5000);
    Serial.println(ESP8266.readString());
}

// logic to disconnect from Wi-Fi
void disconnectWifi()
{
    ESP8266.write("AT+CWQAP\r\n");
    delay(2000);
    Serial.println(ESP8266.readString());
}
```

Now, let's try to understand the sketch. It is already obvious that the sketch is written in a modular way such that it can be used via the Serial Monitor window. Once in the Serial Monitor window, start interacting with the sketch by issuing the numbers 1 to 6 in succession (do not change the order of execution, otherwise the data will not get posted to the cloud), in order to execute the sequence of functions required for sending data to the ThingSpeak IoT Cloud. The `loop()` function is already coded with the following `if` statements to invoke the appropriate function as you enter the number from the Serial Monitor window. If the number 1 is entered in the Serial Monitor window, then the `AT()` function will be invoked. The `AT()` function is used to ping the ESP8266-01 chip with an `AT` command and wait for its response:

```
if (cmd == '1') { AT(); }
```

If the number 2 is entered in the Serial Monitor window, then the `connectWifi()` function will be invoked. This function sends the appropriate AT command to the ESP8266-01 chip for connecting to the specified Wi-Fi SSID:

```
if (cmd == '2') { connectWifi(); }
```

If the number 3 is entered in the Serial Monitor window, then the `CIPSTART()` function will be invoked. This function sends the appropriate `AT` command to the ESP8266-01 chip for starting a new data transmission session:

```
if (cmd == '3') { CIPSTART(); }
```

If the number 4 is entered in the Serial Monitor window, then the `CIPSEND()` function will be invoked. This function sends the appropriate `AT` command to the ESP8266-01 chip for initializing parameters for the data transmission request:

```
if (cmd == '4') { CIPSEND(); }
```

If the number 5 is entered in the Serial Monitor window, then the sendRequest() function will be invoked. This function sends the appropriate AT command to the ESP8266-01 chip for the actual data to be transmitted over the internet:

```
if (cmd == '5') { sendRequest(); }
```

If the number 6 is entered in the Serial Monitor window, then the disconnectWifi() function will be invoked. This function sends the appropriate AT command to the ESP8266-01 chip for disconnecting from the currently connected Wi-Fi network:

```
if (cmd == '6') { disconnectWifi(); }
```

The following Serial Monitor window shows the series of steps that were executed by entering the numbers 1 through 6 in the Serial Monitor window:

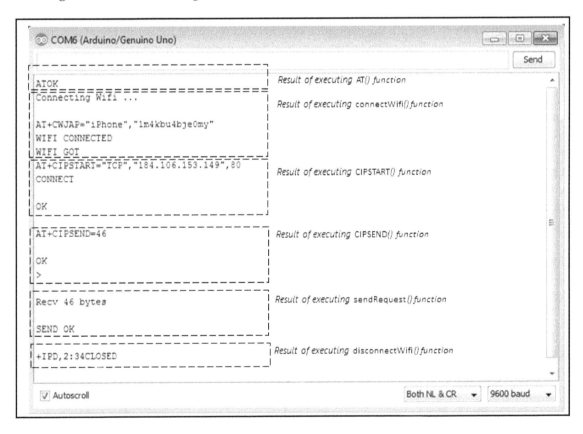

Figure 14: Execution results in the Serial Monitor

Once the desired data has been transmitted to the IoT cloud, then a whole new set of activities can be performed on the cloud platform, depending upon the requirements of the project. However, the most basic of those operations is to be able to fetch the logged data using any application running on automated computer systems. For example, the data logged by a fire detector can immediately trigger a process to alert the Fire Department. In order to fetch the last logged data, all we need to do is use the URLs for fetching data from the IoT cloud. The data fetching URLs can be invoked from any system, be it a C/C++, JAVA or .NET-based application running on computers, mobile phones, or other embedded devices.

Smart retail project inspiration

In the previous sections, we looked at the basics of achieving IoT prototyping with the Arduino platform and an IoT cloud platform. With this basic knowledge, you are encouraged to start exploring more advanced IoT scenarios. As a future inspiration, the following smart retail project idea is being provided for you to try by applying the basic principles that you have learned in this book. After all, the goal of this book has been to show you the light and make you self-reliant with the Arduino platform.

Imagine a large retail store where products are displayed in hundreds of aisles and thousands of racks. Such large warehouse-type retail layouts are common in some countries, usually with furniture sellers. One of the time-consuming tasks that these retail shops face is to keep the price of their displayed inventory matched with the ever changing competitive market rates. For example, the price of a sofa set could be marked at 350 dollars on aisle number 47 rack number 1.

Now let's think from a customer's perspective. Imagine being a potential customer, standing in front of the sofa set; we would naturally search for the prices of that sofa set on the internet. It would not be very surprising to find a similar sofa set that is priced a little lower, maybe at 325 dollars at another store. That is exactly how and when a potential customer would change her mind. The story after this is simple. The customer leaves store A, goes to store B, and purchases the sofa set at 325 dollars.

In order to address these types of lost sale opportunities, the furniture company management decides to lower the prices of the sofa set to 325 dollars, in order to match the competition. Thereafter, all that needs to be done is for someone to change the price label for the sofa set in aisle number 47 rack number 1, which is a 5-10 minute walk (considering the size of the shop floor) from the shop back office. In a localized store, it is still achievable without further loss of customers.

Now, let's appreciate the problem by thinking hyperscale. The furniture seller's management is located centrally, say in Sweden, and they want to dynamically change the product prices for specific price-sensitive products that are displayed across more than 350 stores, in more than 40 countries. The price change should be automatic, near real time, and simultaneous across all company stores.

Given the preceding problem statement, it seems a daunting task that could leave hundreds for shop floor staff scampering all day long, just to keep changing price tags for hundreds of products. An elegant solution to this problem lies in the Internet-of-Things.

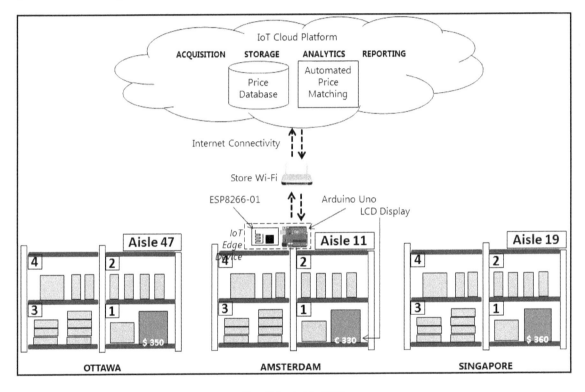

Figure 15: Smart retail with IoT

Referring to the preceding figure, it depicts a basic IoT solution for addressing the furniture seller's problem to matching product prices dynamically on the fly. Remember, since this is an Arduino prototyping book, the stress is on edge device prototyping. However, IoT as a whole encompasses many areas that include edge devices and cloud platform capabilities. Our focus in this section is to be able to comprehend and build the edge device prototype that supports this smart retail IoT solution.

In the preceding solution, the cloud platform takes care of running intelligent program batches to continuously analyze market rates for price-sensitive products. Once a new price has been determined by the cloud job/program, the new product prices are updated in the cloud-hosted database. Immediately after a new price change, all the IoT edge devices attached to the price tag of specific products in the company stores are notified of the new price.

Utilizing the knowledge that we gained in Chapter 6, *Day 4 - Building a Standalone Device*, we can build a smart LCD panel for displaying product prices. For internet connectivity, we can reuse the ESP8266 Wi-Fi chip that we learned in this chapter. Taking the LCD wiring described in Chapter 6, *Day 4 - Building a Standalone Device* simply merge the ESP8266 wiring described in this chapter.

IOT project considerations

IOT projects can be deployed in both sheltered as well as remote weather beaten locations. The following are some considerations while working on IOT projects. As you read through the considerations, you will realize that some of the main problems are power consumption, device security, and weather proofing.

Power source considerations include:

- Avoid batteries (unless unavoidable) as this will increase operational overhead of tracking and replacing. Instead, provide good quality power adapters (AC -> DC 3.3V, 60 - 170 mA) for powering the ESP modules.
- The ESP series Wi-Fi/GSM/GPRS/RF modules need to be powered from a stable battery power source. If electricity mains are not available, then use a rechargeable battery via Solar Panels.
- Battery lifetime enhancement via device/embedded software design: If possible, the Arduino boards and any Wi-Fi/GSM/GPRS/RF chips should be put to sleep mode to conserve energy.

Device enclosure considerations include:

- Devise a proper enclosure for the "Thing". Remember the 3D printing technique discussed in Chapter 6, *Day 4 - Building a Standalone Device*.
- Weather proofing is very important for enclosing remote devices. A suitable weather proof enclosure needs to be procured.

Device security considerations include:

- This is another major area of concern for outdoor locations and has to be dealt with in a conventional manner for ensuring physical security.
- Device identity/global positioning: In order to identify the location of a remote edge device and validate its identity, you will have to attach its identity either while flashing the embedded code, or use an SD card module with a public/private key to be used while communicating with the cloud.

Try the following

This being the last chapter, let's try to complete the following exercise before concluding this journey:

- Refer to the information provided in `Chapter 5`, *Day 3 - Building a Compound Device*, and build a connected smoke detector. The connected smoke detector should send data to the IoT cloud whenever smoke is detected. Reuse the ESP8666-01 chip, the ThingSpeak IoT cloud, and the relevant sections from the sketch in this chapter for posting data to the cloud.

Things to remember

Remember the following important points about what we learned in this chapter:

- An IoT solution encompasses edge devices as well as IoT cloud platforms. We use the Arduino platform for rapidly prototyping edge devices for supporting IoT solutions.
- Some of the popular internet connectivity devices that can be used with the Arduino are: GPRS/GSM modules, ESP8266 series Wi-Fi chips, and Ethernet shields.
- A device channel is a pre-configured software defined endpoint in an IoT cloud for data exchange with an IoT edge device.
- Data can be exchanged with IoT cloud platforms using HTTP-based URLs. The correct format and URL structure will vary depending upon the IoT cloud.
- Read and write API keys are used to protect data exchange between the cloud and the edge device.

- Sometimes IoT cloud vendors may provide their custom libraries for various edge devices, for easily connecting and exchanging data with the cloud platform. Cloud platform specific documentation must be consulted in such cases.
- Just in case the ESP8266 chip does not respond to the Arduino sketch or AT commands properly, then the power supply may not be enough. Try using a separate battery for the setup. Remember to use a voltage regulator that steps down the voltage to 3.3 volts before supplying the ESP8266 chip.

Summary

In this chapter, we were introduced to the world of the Internet of Things. We learned how to use the ESP8266-01 Wi-Fi chip for transmitted device data to ThingSpeak (an IoT cloud platform) with the Arduino platform.

With this we come to the conclusion of our crash course for learning Arduino platform-based device prototyping. As a next step, the reader is encouraged to start thinking and building more device prototypes based on each chapter in this book. Start with simple sensors and then move on to battery-powered compound devices, followed by building more and more wireless devices. Having crash landed into the world of Arduino, the journey begins from here.

Index

2

2.4 GHz ISM band 210

3

3D printed project boxes
  creating 141

5

5 volt tolerant peripheral devices 96

A

AC powered devices
  relays, using 168, 169, 170
actuators 146
analog input and output 35, 36, 37
Arduino board
  D0 (Rx) 234
  D1 (Tx) 234
  setting up 23, 24
  URL, for downloading 23
Arduino components
  analog I/O pins 20
  DC IN jack 20
  digital I/O pins 20
  flash memory 20
  interrupt pins 20
  Power pin (3.3V output) 20
  Power pin (5V output) 20
  Power pin (Ground) 20
  Power supply pin (input) 20
  USB B port 20
Arduino Infrared library
  URL 188
  using 188
Arduino interrupts
  URL 157

using 156
Arduino platform
  about 18, 19, 20
  prototyping 21, 22
Arduino program
  execution 24, 25, 26
  structure 24, 25, 26
Arduino projects
  controlling 205, 206
Arduino Uno
  about 19
  GSM module, interfacing 231
  HC-05 Bluetooth module, connecting 221, 222
  reference 19
AT commands
  about 230
  AT 230
  AT+CMGF=1 230
  AT+CMGS="1234567890" 230
  ATD1234567890; 230
  ATH 230
ATmega328P microcontroller 19

B

basic sensor components
  about 66
  photo resistor (LDR), using 71, 73, 75
  photodiode LED, using 66, 67, 69, 71
Bluetooth communications
  about 219
  HC-05 Bluetooth module, using 220
Bluetooth Low Energy (BLE) module 220
breadboard circuit 43

C

common grounding 43
compound device 96

Consumer Infrared (CIR) 185

D

datasheet 75
DC motor
  about 146
  prototype, building 147, 149, 150
  sketch, designing 150, 151
  speed control sketch, designing 154, 155, 156
  speed, controlling with PWM method 151, 153
  usage considerations 146
DHT11
  about 76
  datasheet, URL 77
digital input and output 33, 34
digital signals
  receiving 103
  sending 103
diodes
  about 39
  using 55, 56
distance measurement device
  building 129, 130, 131
  circuit, building 132, 133, 135
  operating 138
  power switch, using 139, 141
  project enclosure 139, 141
  sketch 135, 137, 138

E

edge device channel
  configuring 250, 251, 252, 254, 256
EEPROM
  references 108
electronic signage industry 46
external power supply
  options 125, 126

F

fire alarm 96
forest fire early warning system
  building 237, 239
fritzing 42

G

General Packet Radio Services (GPRS) 228
General Purpose Input Output (GPIO) pins 19, 21
Global System for Mobile (GSM) 228
GSM module
  about 228, 229
  interfacing, with Arduino Uno 231
  sketch, writing 232, 234, 235, 237
  using 227

H

Hayes command set 230
HC-05 Bluetooth module
  communicating, with prototype 224
  connecting, to Arduino Uno 221, 222
  sketch, writing 222
  using 220
HC-SR04 Ultrasonic sensor 34
HC-SR04
  about 130
  Echo pin 131
  Gnd pin 131
  Trig pin 131
  Vcc pin 131

I

in-built function, sketch
  about 31
  analogRead(PIN-NUMBER) 32
  analogWrite(PIN-NUMBER, SIGNAL-VALUE) 32
  delay(MILLI-SECONDS) 32
  delayMicroseconds(MICRO-SECONDS) 32
  digitalRead(PIN-NUMBER) 32
  digitalWrite(PIN-NUMBER, SIGNAL-LEVEL) 32
  isnan(VALUE) 32
  millis() 33
  pinMode(PIN-NUMBER, I/O-MODE) 31
  pulseIn(PIN-NUMBER, LOGIC-LEVEL) 33
  Serial.begin(BAUD-RATE) 32
  Serial.print("MESSAGE") 32
  Serial.println("MESSAGE") 32
  tone(PIN-NUMBER, FREQUENCY, MILLI-SECONDS) 33
Industrial, Scientific and Medical (ISM) community

210

Infrared (IR) communications
  about 184
  frequency 185
  protocols 185
Infrared receiver device
  Arduino Infrared library 188, 190, 192, 193
  building 187
  SM0038 IR receiver, using 200
  TSOPseries IR receivers, using 194, 196, 197, 199
Infrared transmitter device
  building 201
  IR transmitter LED, using 201, 203, 204
Integrated Circuits (ICs) 19
Integrated Development Environment (IDE) 23
integrated sensor modules
  program output, viewing 84
  sensor connection, determining 79, 80
  sensor interfacing sketch 82
  sensor-specific Arduino library, installing 81
  soil moisture sensor module, using without
    Arduino library 86
  temperature sensor module, using with Arduino
    library 76
  using 76
Internet of Things (IoT) 228, 243, 244
IoT cloud platforms
  about 247
  configuration 249
  edge device channel, configuring 250, 251, 252, 254, 256
  fundamental services 247
  registration 249
  selecting 247
IoT edge devices
  about 245
  building 257, 259
  fundamental parts 247
  setting up 257
  sketch, loading 259, 261, 263
  usage 245
IoT projects
  considerations 265
IR library

URL 190

J

JSON (JavaScript Object Notation)
  about 256
  online resource 256

L

Light Dependent Resistor (LDR) 71, 75
Light Emitting Diodes (LEDs)
  about 39
  with push button 59, 60, 61, 62
local storage
  with SD card modules 109, 110, 112, 113, 117, 118, 119

M

Machine-to-Machine (M2M) communications 210
maintained switch 140
maniacbugRF library
  URL 214
Messaging Queue Telemetry Transport (MQTT)
  protocol 245
micro SD card 109
momentary switch 139
MQ2 series gas detector module 96

N

network (NTW) LED 231
Nordic Semiconductor 211
nRF24L01 transceiver module
  open source RF library for Arduino, downloading 214
  radio frequency signals, receiving 217, 218
  radio frequency waves, transmitting 215, 216
  RF transmitter-receiver pair, testing 219
  using 211
  wiring, with Arduino 212, 214

O

Ohm's Law 41
open source RF library
  downloading, for Arduino 214

P

photo resistor (LDR)
  using 71, 73, 75
photodiode LED
  using 66, 67, 69, 71
Piezo Buzzer
  using 39, 47, 48, 96
power source capacity
  determining 127, 128
power switch
  maintained switch 139
  momentary switch 139
  using 141
project enclosure
  about 140
  references 141
  using 141
Pulse Width Modulation (PWM) 20, 151
push button
  used, for Light Emitting Diodes (LEDs) 59, 60,
    61, 62
PWM method
  used, for controlling speed of DC motor 151, 153

R

radio frequency (RF) 209
radio frequency device
  Bluetooth communications 219
  building 210
  nRF24L01 transceiver module, using 211
radio frequency signals
  receiving 217, 218
radio frequency waves
  transmitting 215, 216
rectifier diodes 55
Reduced Instruction Set (RISC) 19
relay device
  Common (C) 168
  Normally Closed (NC) 169
  Normally Open (NO) port 168
relays
  using, with AC powered devices 168, 169, 170
remote control
  hacking into 186

remote-controlled reconnaissance boat
  building 164, 165
resistors 39
  using 40, 41, 42, 43, 45, 46
RF transmitter-receiver pair
  testing 219
room lights
  automating 180, 181

S

S-Link 186
SD card module, pins
  CS/SS (Chip Select/Slave Select) 110
  GND 111
  MISO (Master In Slave Out) 111
  MOSI (Master Out Slave In) 111
  SCK/CLK (System Clock) 111
  VCC 111
SD card modules
  about 108
  used, for local storage 109, 110, 112, 113, 117,
    118, 119
second generation (2G) 228
sensor components
  basic sensor components 66
  integrated sensor modules 66
  types 66
servo motor
  control circuit 159
  control sketch, designing 161, 162, 163
  interfacing with 158
SIM800 chip 229
Single Pole Single Throw (SPST) 140
sketch
  about 24
  compiling 28, 30, 31
  executing 28, 30, 31
  exploring 26, 27, 28
  in-built function 31
  loading 28, 30, 31
SM0038 IR receiver
  5V 200
  GND 200
  Out 200
  using 200

smart irrigation systems
  building 92
smart retail project 263, 264
smoke detector device
  about 96
  analog I/O method, examining 104
  building 96, 98
  digital I/O method 98, 99, 100
  sketch, for analog I/O method 105, 107, 108
  sketch, using 100, 102, 103
soil moisture sensor module
  circuit 88, 89
  sketch 89, 91
  using, without Arduino library 86
sound activated light bulb controller
  prototype 177, 179
  simulating 171, 172, 174
  sketch, designing 174, 176
stackable GSM shield
  reference 229
standalone devices 124
switching 51

T

temperature sensor module
  sensor connection, determining 78
  sensor module datasheets 77
  using, with Arduino library 76
ThingSpeak
  about 248
  URL 248, 249
three LED project 40
tinkercad
  URL 141
transceivers 206
transistors 39
  using 51, 53, 54, 55
TSOPseries IR receivers
  5V 194
  GND 194
  Out 194
  using 194, 196, 198, 199

U

URL formats
  reference 254, 256

V

voltage division technique 221

www.ingramcontent.com/pod-product-compliance
Lightning Source LLC
Chambersburg PA
CBHW060522060326
40690CB00017B/3356